SpringerBriefs in Geography

SpringerBriefs in Geography presents concise summaries of cutting-edge research and practical applications across the fields of physical, environmental and human geography. It publishes compact refereed monographs under the editorial supervision of an international advisory board with the aim to publish 8 to 12 weeks after acceptance. Volumes are compact, 50 to 125 pages, with a clear focus. The series covers a range of content from professional to academic such as: timely reports of state-of-the art analytical techniques, bridges between new research results, snapshots of hot and/or emerging topics, elaborated thesis, literature reviews, and in-depth case studies.

The scope of the series spans the entire field of geography, with a view to significantly advance research. The character of the series is international and multidisciplinary and will include research areas such as: GIS/cartography, remote sensing, geographical education, geospatial analysis, techniques and modeling, landscape/regional and urban planning, economic geography, housing and the built environment, and quantitative geography. Volumes in this series may analyze past, present and/or future trends, as well as their determinants and consequences. Both solicited and unsolicited manuscripts are considered for publication in this series.

SpringerBriefs in Geography will be of interest to a wide range of individuals with interests in physical, environmental and human geography as well as for researchers from allied disciplines.

Maciej J. Nowak · Andrei Mitrea · Gunta Lukstiņa · Alexandru-Ionuţ Petrişor · Krisztina Filepné Kovács · Velislava Simeonova · Pavel Yanchev · Evelin Jürgenson · Kätlin Põdra · Vít Řezáč · Kristina Mikalauskaite · Birute Pranevičienė · Zuzana Ladzianska · Martin Baloga

Spatial Planning Systems in Central and Eastern European Countries

Review and Comparison of Selected Issues

Springer

Authors
See next page

ISSN 2211-4165 ISSN 2211-4173 (electronic)
SpringerBriefs in Geography
ISBN 978-3-031-42721-3 ISBN 978-3-031-42722-0 (eBook)
https://doi.org/10.1007/978-3-031-42722-0

This Springer imprint is published by the registered company Springer Nature Switzerland AG
The registered company address is: Gewerbestrasse 11, 6330 Cham, Switzerland

Paper in this product is recyclable.

Maciej J. Nowak
Real Estate Department
West Pomeranian University of Technology
Szczecin, Poland

Gunta Lukstiņa
Faculty of Geography and Earth Sciences
University of Latvia
Riga, Latvia

Krisztina Filepné Kovács
Department of Landscape planning
and Regional Development
Hungarian University of Agriculture and Life
Sciences
Budapest, Hungary

Pavel Yanchev
Brussels, Belgium

Kätlin Põdra
Forest and Land Management and Wood
Processing Technologies, Institute of Forestry
and Engineering
Estonian University of Life Sciences
Tartu, Estonia

Kristina Mikalauskaite
Law and Law Enforcement Institute,
Academy of Public Security
Mykolas Romeris University
Kaunas, Lithuania

Zuzana Ladzianska
SPECTRA EU Centre of Excellence
Slovak University of Technology
Bratislava, Slovakia

Andrei Mitrea
Department of Urban Planning and Territorial
Development, School of Urban Planning
"Ion Mincu" University of Architecture
and Urban Planning
Bucharest, Romania

Alexandru-Ionuţ Petrişor
Doctoral School of Urban Planning
"Ion Mincu" University of Architecture
and Urban Planning
Bucharest, Romania

Department of Architecture, Faculty
of Architecture and Urban Planning
Technical University of Moldova
Chişinău, Moldova

National Institute for Research
and Development in Constructions, Urbanism
and Sustainable Spatial Development
URBAN-INCERC
Bucharest, Romania

National Institute for Research
and Development in Tourism
Bucharest, Romania

Velislava Simeonova
Faculty of Geography and History,
Department of Geography
University of Barcelona
Barcelona, Spain

Evelin Jürgenson
Forest and Land Management and Wood
Processing Technologies, Institute of Forestry
and Engineering
Estonian University of Life Sciences
Tartu, Estonia

Vít Řezáč
Spatial Planning Department, Faculty
of Architecture
Czech Technical University in Prague
Prague, Czech Republic

Birute Pranevičienė
Law and Law Enforcement Institute,
Academy of Public Security
Mykolas Romeris University
Kaunas, Lithuania

Martin Baloga
SPECTRA EU Centre of Excellence
Slovak University of Technology
Bratislava, Slovakia

Introduction

Comparisons of national spatial planning systems have long taken centre stage in the dedicated literature. However, a definitive agreement on how to compare and contrast different countries is not in sight at the time of writing (2023). Furthermore, planning systems across Central and Eastern European countries—the New Member States of the European Union—have largely been neglected. It is against this background that the main aim of this book is to compare key aspects of spatial planning systems in Romania, Bulgaria, Hungary, the Czech Republic, Slovakia, Lithuania, Latvia, Estonia, and Poland.

Hence, the first part of the monograph studies the classification of their spatial planning systems, focusing on the different national interpretations of spatial planning concepts, as well as on the institutional constellations that engage in spatial planning. The second part goes deeper into the realm of key planning instruments, looking closely at spatial plans at the local, regional, and national levels, as well as select policy documents dealing with spatial development.

Going into a bit more detail, Chap. 1 presents the main ways of classifying and comparing national spatial planning systems. Their salient features and the barriers they face in implementing concepts and regulations are the main points of comparison here. In addition, we reveal the different understanding of key planning concepts and values are understood across CEE countries and highlight terminological discrepancies. The chapter includes relevant references for the current international discussion.

Chapter 2 begins by outlining the common historical and spatial context of CEE countries. This is followed by a brief description of the key spatial planning features and instruments in each country. Depictions appear in tabular form, accompanied by a brief commentary. We have reviewed spatial plans at different levels (their role, legal significance, obligatory nature, and content), as well as development documents with a spatial dimension. Attention was also paid to the environmental and nature conservation dimensions of spatial planning. This chapter provides quick orientation and insight into the basic features of CEE planning systems.

In Chap. 3, an extensive comparative analysis is carried out. Both similarities and differences appear highlighted, with authors comparing national solutions in their respective countries, diagnosing their strengths and weaknesses. They will therefore refer more extensively to the problems found within spatial planning systems, followed by more in-depth case studies of select instruments. The issue of integrating the environmental and natural dimensions with spatial planning plays a special role here.

And finally, the summary contains a synthesis of reflections, as well as future research directions.

Summary

When perusing the literature, the objectives of spatial planning appear very broad and ambitious (Koresava and Konvitz 2001). The essence of spatial planning involves multidimensionality and the need to integrate different scales and fields (Lopez and Trigal 2015). For this reason, it seems crucial to look closer at different planning instruments and the links between them. This is where spatial planning becomes operational, and the performance of planning systems becomes visible.

Against this background, the analyses carried out in the monograph probe the discrepancies between the strategic dimension of spatial planning and its regulatory dimension. Our main aim was to understand how planning concepts are translated into regulations. In strategic spatial planning, objectives are first worked out and then defined. Afterwards, an arduous process of translation begins, rendering them fit for regulations. Looking across countries in Central and Eastern Europe, we see different solutions of transposing planning concepts into operative regulations. The crux of the matter here is to have viable regulations that limit blatant and insidious transgressions, without stifling development. We have found that:

- Strategic or conceptual guidelines are sometimes too vague to be relevant from a legal perspective.
- Sometimes, these guidelines are not implemented at all (e.g., by not enacting local spatial plans, thereby confirming the views expressed by Kahrin and Stevan 2018).
- Excessively detailed legal provisions are often attached to these guidelines.

These results confirm that the role of law in spatial planning systems should become the subject of broader theoretical study (Needham et al. 2019). Clearly, one important aspect here is the detailed approach to spatial planning legislation (Savini and Aalberts 2016, Buitelaar 2012).

There is no doubt that, in plan-based systems, binding provisions should at least encompass zoning and substantiated development restrictions. What is debatable, however, is how to include more detailed development parameters into plans. This should depend on the specifics of the individual site, the needs involved, and the planning culture in the country concerned.

The concept of planning flexibility should also be adapted to the circumstances of each country (Muñoz-Gielen and Tasan-Kok 2010). When viewed from the perspective of Central and Eastern European (CEE) countries, flexibility in planning should be paramount, as it might solve many of the problems mentioned above. On the other hand, this flexibility should not be abused. It is here that integrated development planning becomes important, as it integrates different institutional perspectives. It can produce a variety of solutions: either an attempt to fit strategic/conceptual and regulatory issues into a single document or to produce separate documents, aimed at integrating development policies.

Of course, there is also a risk, especially in the latter case, of not applying these provisions in practice. De Vries (2015) pointed out that the similarity of provisions does not determine the actual planning practice in a country. This observation also applies to the CEE countries studied here: there are strong analogies between legal solutions and, implicitly, similarly diagnosed problems. However, specific solutions usually differ across planning systems.

Hence, this book provides evidence for the similarity of problems confronting CEE planning systems, and for the barriers they face. It thereby contributes to a wider international discussion on spatial planning problems and systems, as it provides a much-needed international perspective that has been hitherto neglected. Essentially, it complements the study by Nadin et al. (2018) that reviewed planning systems in 39 European countries and highlights the salient differences in understanding and employing concepts such as spatial planning and spatial plans, spatial development, and land use across CEE countries.

In addition, the case studies included in the monograph show in detail the individual problems that may arise in different spatial planning systems. They may be related to weaknesses in individual regulations, but very often, their actual cause lies in an ingrained misunderstanding of the challenges and needs of spatial planning.

Against this background, the results presented in the following chapters serve as a platform for discussing the following topics:

- The normative framework of spatial planning,
- The role of courts in spatial planning,
- The strategic dimension of spatial planning across CEE countries,
- Roots of the planning discourse in CEE countries: national practices, European financing, or European discourse,
- Articulating planning instruments with planning practices,
- Implementing sectoral objectives through specific spatial planning instruments.

References

Buitelaar E (2012) The fraught relationship between planning and regulation. Land use plans and the conflicts in dealing with uncertainty. In: Needham B, Hartmann T (eds) Planning by law and property rights reconsidered. Routledge, New York, pp 207–218

Koresawa A, Konvitz J (2001) Towards a new role for spatial planning. OECD Publishing, Paris. 10.1787/9789264189928-en

López, Trigal L (eds) (2015) Diccionario de geografía aplicada y profesional: terminología de análisis, Planificación y Gestión del Territorio. Universidad de León, Leon
Muñoz-Gielen D, Tasan-Kok T (2010) Flexibility in planning and the consequences for public value capturing in UK, Spain and the Netherlands. Eur Plan Stud 18(7):1097–1131. 10.1080/096543 11003744191
Nadin V, Fernández-Maldonado A, Zonneveld W et al. (2018) COMPASS—comparative analysis of territorial governance and spatial planning systems in Europe applied research 2016–2018 final report. European Commission ESPON, Luxembourg
Needham B, Buitelaar E, Hartmann T (2019) Planning law and economics: the rules we make for using land, 2nd edn. RTPI Library Series, Routledge, New York
Savini F, Aalbers MB (2016) The de-contextualisation of land use planning through financialisation: Urban redevelopment in Milan. Eur Urban Reg Stud 23(4):878–894. 10.1177/096977641558 5887
de Vries J (2015) Planning and culture unfolded: the cases of flanders and the Netherlands. Eur Plan Stud 23:112148–112164. 10.1080/09654313.2015.1018406

Contents

Chapter 1
Barriers in Comparing Spatial Planning Systems Across Different Countries

Abstract The aim of the chapter is to compare approaches to the basic planning concepts and issues across CEE countries that are also member states of the European Union (EU). Based on their previous cooperation, the authors concluded that the following issues are important and worthwhile:

- Defining spatial planning and its objectives,
- Understanding spatial plans and their related concepts,
- Grasping the relationship between spatial planning and development policy, and the barriers that arise between them.

Defining spatial planning allows for broader thematic comparisons. Spatial planning objectives, on the other hand, are the key reference point for the individual spatial planning instruments, both in terms of their creation and their evaluation. In addition to reflections resulting from observations of their own systems, the authors present reflections based on preliminary comparisons of solutions provided in CEE countries. Results stem from both ongoing analyses and previous joint research.

Keywords Spatial planning · CEE countries · Spatial plans · Values in spatial planning

1.1 Introduction

There is undoubtedly a need for an in-depth discussion comparing the spatial planning systems of different countries. Apart from its scientific appeal, such a debate might prove one of the best ways to improve specific planning solutions within these countries.

However, there is a considerable risk of simplification. For example, a comparison between two specific laws in two (or more) countries will yield only moderately rewarding results. Hence, it becomes necessary to consider the various planning traditions, their specific solutions, as well as the day-to-day planning practices of individual countries. Differing features of national planning systems make this task considerably more difficult. This also applies to geographically close countries, for

M. J. Nowak et al., *Spatial Planning Systems in Central and Eastern European Countries*, SpringerBriefs in Geography, https://doi.org/10.1007/978-3-031-42722-0_1

example the countries of Central and Eastern Europe (CEE). Here, it becomes important to distinguish between common features, which are, by and large, noticeable, and differences that are still not fully identified.

Approaches to spatial planning are very different. Different views occur within different scientific disciplines and in different countries, especially when discussing legal systems. Moreover, even within a single discipline or within a single country, there are differences in terminology, which arise for various reasons. From the perspective of a scientific discussion, differing approaches are always good, as they allow for comparisons. However, comparing the planning systems of different countries does require a common approach, which we have developed later in the book.

Against this background, the aim of the first chapter is to compare approaches to the basic planning concepts and issues across CEE countries that are also member states of the European Union (EU). Based on their previous cooperation, the authors concluded that the following issues are important and worthwhile:

- Defining spatial planning and its objectives,
- Understanding spatial plans and their related concepts,
- Grasping the relationship between spatial planning and development policy, and the barriers that arise between them.

Defining spatial planning allows for broader thematic comparisons. Spatial planning objectives, on the other hand, are the key reference point for the individual spatial planning instruments, both in terms of their creation and their evaluation. It is equally important to define the relationship between spatial planning and development policy (development planning): Firstly, this is a question regarding the legal consequences of planning standards. For example, current scientific discourse is very often concerned whether certain (detailed) legal norms aid or block development (Savini and Salet 2016). Secondly, this relationship also determines how a particular system works. Hence, development-based systems understand a significant part of spatial planning tasks differently from plan-based systems (Muñoz Gielen and Tasan-Kok 2010; Tasan-Kok 2008). In addition, the historical experience of CEE countries shows how different the relationship between spatial planning and socio-economic planning can be. It is therefore highly important to analyse such problems from this perspective.

A separate part of this chapter discusses the directions for classifying spatial planning systems. This debate is ongoing and very extensive, with the authors wishing to contribute with their opinions and observations on the subject. In addition to reflections resulting from observations of their own systems, the authors present reflections based on preliminary comparisons of solutions provided in CEE countries. Results stem from both ongoing analyses and previous joint research (Nowak et al. 2022a).

1.2 Spatial Planning—Meaning and Approaches

1.2.1 Literature Review

The concept of spatial planning appears very often in the literature. Nadin et al. (2018) understand it as the process of organising territory and land use, as well as managing competing interests with the aim to guide development ensuring environmental protection, land conservation and the involvement of local communities.

These authors also emphasise that the term 'spatial planning' implies an active collaboration between government, market, and the civil society, aimed at coordinating decision-making that affects the quality of places and their development. This definition and understanding currently serve as one of the main reference points for the academic discourse, but different spatial planning aspects can (and should) be considered.

At the same time, the literature notes that spatial planning is often replaced by other concepts, such as 'spatial development' or 'strategic planning' (Luukkonen 2011). However, when viewed from a purely theoretical perspective, these concepts have different meanings. It is worth noting that current approaches to spatial planning (and the ways of understanding it) stem from three different strands of thought (Healey 1997): economic planning, with its aim is to manage the productive forces of nations and regions, urban development management, which promotes health, economy, comfort, and beauty in the urban environment (Abercrombie 1933; Keeble 1952; Adams 1994), and, finally, public administration management and policy analysis. Spatial planning thereby combines three different traditions: Thus, it encompasses the relationship with economic planning, the protection of architectural (and other) values of a given area, and the link to public policies.

Healey (1997, 2004) defined spatial planning as a set of management practices for developing and implementing strategies, plans, policies, and projects, as well as for regulating the location, timing, and form of development. She also indicates that this term refers to conscious collective efforts to re-imagine a city, an urban region or a wider territory and their translation into priorities for infrastructure investment and land use policies. Against this background, the term 'spatial' essentially focuses on the 'place of things'. It refers to things that are both static and in motion. At the same time, it includes the protection of special 'places and objects', while at the same time defining the interaction between activities and networks in each area, as well as the intersections and nodes proper. Such an approach becomes particularly clear when answering the general question of the role of spatial planning in relation to a specific place or territory.

This connects easily with the view that 'spatial planning' refers to the management of land and property, as well as to the promotion of preferred forms of spatial and urban development through strategies that integrate the spatial dimensions of sectoral policies (Cullingworth and Nadin 2006). In systemic terms, following López and Trigal (2015), spatial planning is the foundation for integrating the different elements and determinants of spatial dynamics. In particular, it is about organising and structuring the relationship between society, territory, and economy, necessarily taking into account a medium and long-term vision. Tewdwr-Jones (2012) stresses the role of the political economy and institutional restructuring, where planning is not a solution but only a means for managing spatial processes. According to him, the term 'spatial planning' includes describing 'the many facets of planning practices that provide proactive opportunities for managing change, involving policy making, policy integration, community participation, agency participation, and development management' (Tewdwr-Jones and Allmendinger 2006). Hence, linking spatial planning, especially at the local level, to the concept of governance is also useful in this view (Virtudes 2015). Koresawa and Konvitz (2001) consider spatial planning as a tool for balancing social, economic, and environmental needs, as well as providing an institutional, technical, and political framework for managing the territorial dimension of sustainable development, while at the same time protecting the rational organisation of spatial activities.

An additional approach to the term 'strategic spatial planning' can also be invoked (Albrechts 2015), which is not always singled out independently. For the purposes of further discussion, however, it is worth pointing out that Hutter and Wiechmann (2021) and Healey understand it as a social process in which a range of people in different institutional relationships and positions come together to design processes for plan-making and to develop content and strategies for managing spatial change. Hersperger et al. (2018a) point out that the dominant view of strategic spatial planning aligns very well with functional city regions and a strong focus on strategic mission, often 20–50 years into the future. At the same time, strategic spatial planning comes across as difficult to capture intuitively, given the fact that is multidimensional, embedded in socio-political and institutional complexity, and highly context-dependent. According to these authors, current planning debates mainly focus on the outcomes of the strategic planning process, while largely neglecting the impact that strategic spatial plans can have on urban transformation (Hersperger et al. 2018b).

Against this background, we can deduce that:

- At least two levels of spatial planning can be distinguished: The first concerns the development of a broader concept to guide the development and role of specific areas. This includes defining the relationship with economic planning, reconciling conflicting interests, protecting certain spatial values, and translating the critical approach to land into tangible measures. The second level concerns the realm of implementation, which usually boils down to specific land use guidelines and regulations. The challenge here is to maintain compatibility between the two levels. For want of a better name, we can call these the 'conceptual level' and the 'implementation level'.
- The multidimensionality of spatial planning also deserves an emphasis because it includes many different dimensions: environmental, natural, social, and economic. In different countries, the scope of approaching these dimensions varies significantly, especially in the way in which they are integrated into spatial planning activities.

- The concept of 'strategic spatial planning' is not the same as the 'conceptual level' identified above, although it is strongly connected to it and is an essential element of it.

1.2.2 The Concept of Spatial Planning in Legal Acts and the Positions of Public Authorities Across CEE Countries

The legal and organisational orders of individual countries contain the definition of the approach to spatial planning in the sphere of a legal or official (e.g., in national development documents) declaration. The normative frameworks of individual CEE countries offer both the definitions of 'spatial planning' and their approaches to it. Note, however, that a legal definition will always be simpler and more general than a theoretical interpretation. At the same time, approaches to spatial planning in normative documents remain descriptive.

Hence, in each country, these two issues are dealt with in a slightly different manner. It is also important to remember that 'spatial planning' is a term that is strongly adapted to specific national and linguistic contexts (Dühr et al. 2010). Therefore, individual cases have been analysed and synthesised. The main aim is to provide a broader factual basis for the ensuing discussion.

Lithuanian legislation indicates that territorial planning (it is also referred to there as 'spatial planning') means a process aimed at sustainable territorial development, prioritising land use, environmental protection, public health, and cultural heritage. It should therefore include the creation of residential and productive areas, as well as engineering and social infrastructure systems. Among the many objectives of spatial planning, we can highlight the following:

- Facilitating a balanced territorial development of the country,
- Implementing a coherent functional and spatial policy,
- Devising a comprehensive solution to a wide range of problems, including the challenges of climate change,
- Assisting the rational use and renewal of natural resources,
- Balancing the interests of individuals and the state.

In Latvia, legislator in the Act defines the purpose of spatial development planning—to ensure that the development of a territory is planned in a manner which would raise the quality of the living environment, ensure sustainable, effective, and rational use of territories and other resources, as well as targeted and balanced development of economy. Among others, there are spatial development planning principles:

- Principle of equal opportunities—sectoral and territorial interests, private and public interests are assessed in interconnection to promote sustainable development of the relevant territory.

- Principle of integrated approach—economic, cultural, social, and environmental aspects are harmonised, sectoral interests are coordinated.
- Principle of diversity—development of a territory is planned by taking into account the diversity of natural, cultural environment, human and material resources and economic activities (Act).

In Estonia, spatial planning means the conscious shaping of the external environment, creating conditions for long-term sustainable spatial development, which is democratic and considers the needs and interests of all members within the society. In addition, it formulates conditions for spatial development and the success of a high-quality living and development environment, by promoting environmentally friendly and economically, culturally, and socially sustainable development. Therefore, the main task of spatial planning is to agree on the principles and conditions of spatial development within an area (Act 1996).

Spatial planning in Hungary is strongly divided between land use planning and spatial development planning. Among the objectives of spatial planning in the country are:

- Strengthening territorial cohesion in all regions of the country,
- Promoting social and economic growth,
- Shaping a spatial structure that meets social, economic, and environmental aims,
- Reducing disparities between rural and urban areas (Jürgenson et al. 2017).

Romanian law distinguishes between regional planning and urban (town planning). Both categories are defined as complex activities of a general interest, contributing to balanced spatial development, protection of natural and built heritage, and improvement of living conditions. They are intended to ensure territorial cohesion at the regional, national, and European levels and are oriented towards land management.

Regional planning is thus associated with giving sectoral policies a spatial expression, while urban planning is more concerned with land use and land cover within basic administrative units, as well as the protection of biodiversity, ecological continuity, or ensuring public safety and health. The main aim of regional planning is to ensure a balanced, coherent, and sustainable development of the national territory, to increase economic, social, and territorial cohesion, and to coordinate sectoral policies. The objective of urban planning is the comprehensive development of urban and rural municipalities through the implementation of short-term, medium-term, and long-term development strategies. Hence, among the objectives of regional planning are:

- Sustainable economic and social development of the country's areas and regions, considering their specific characteristics,
- Improving the quality of life,
- The rational use of land, including the prevention of urban sprawl into agricultural areas.

As for the aims of urban planning, the objectives mentioned above are complemented by the improvement in the quality of life using territorial and community resources, as well as by ensuring the architectural and functional quality of the built and natural environment in rural and urban municipalities.

The Czech legislator indicates that the purpose of spatial planning is to create the prerequisites for the construction and sustainable development of an area, consisting in a balanced combination of conditions conducive to the environment, economic development, and community cohesion of that area's inhabitants. Spatial planning thus seeks to provide preconditions for sustainable development, aiming to achieve some level of harmony between public and private priorities, while considering the social and economic potential of development. In addition, spatial planning needs to protect and develop the natural, and cultural values of the area, including urban planning, architectural, archaeological heritage, and landscape, reducing the threat of environmental and natural disasters. Among the specific tasks of spatial planning, we find:

- Establishing and assessing the status of the area,
- Defining a development concept for a given area,
- Examining and assessing the need for change in among public priorities and their implementation,
- Establishing urban, architectural, and aesthetic requirements for the use and spatial development of an area and its changes (usually about location and building layouts),
- Creating conditions for reducing the risk of environmental and natural disasters, removing, or at least diminishing the effects of rapid economic change, and spending public funds sparingly on changes in the area (Act 2006; Maier 2014).

The main instruments of territorial planning in the Slovak Republic are the territorial, so-called land use plans. Territorial planning, in accordance with Slovak legislation, is primarily an instrument of the local government. The so-called mandatory regulations of a higher-level land use plan for the functional and spatial organisation of a given territory are binding for all lower-level plans ('top-down' principle). According to the Act 50/1976 Coll. on Land-use Planning and Building Order, (Act 1976; Ladzianska et al. 2019), the land use planning documentation addresses the spatial arrangement of the land use system, harmonises the interests and activities affecting land development, environment, and ecological stability, while at the same time establishing the directions for the spatial arrangement and functional use of land. Land-use documentations are drafted at the national and at the regional level, as well as for municipalities and for parts of them.

Bulgarian spatial planning employs the term 'territorial planning'. Within this framing, the 'territory' is understood as something to be rationalised and organised. However, the current Spatial Planning Act does not define spatial planning, focusing primarily on procedural issues. Nevertheless, based on its provisions, we can assume that spatial planning should care for favourable conditions for work and leisure, as well as to ensure sustainable development. In addition, based on the literature review of the Bulgarian planning system, we can surmise that spatial planning must, above all, reduce spatial disparities and serve as a framework for strategic development planning (Campesino Fernández 2020).

In Poland, the Act on Spatial Planning and Development does not define the concept of 'spatial planning'. Neither does it explicitly define the objectives of spatial planning. Nevertheless, it indicates what public authorities should consider when engaging in spatial planning:

- Spatial order requirements, including urban planning and architecture,
- Architectural and landscape values,
- Environmental protection requirements,
- Requirements for the protection of cultural heritage and historical monuments,
- Health protection requirements and the safety of persons and property,
- The economic value of space,
- Property rights,
- The needs of national defence and security,
- Public interest needs (Act 2003).

The concept of spatial order requires special attention, as it is understood by the legislator as a land structure that creates a harmonious whole and considers in an orderly manner all functional, socio-economic, environmental, cultural, as well as compositional and aesthetic requirements. This definition is quite broad, especially when viewed from the urban planning and implementation perspective. We can deduce that spatial order is a vision of the ideal development or adjustment of space. This vision will not be fully realised, but one should get as close to it as possible.

Therefore, the expressions 'protection of spatial order' and 'shaping of spatial order' are quite appropriate. In the first case, the aim is to protect the existing valuable composition, aesthetic, environmental, natural, and cultural values of an area. Such protection may first be related to urban development pressure. On the other hand, 'shaping of spatial order' has a broader meaning and may be identified not only with the protection of existing spatial values, but also with appropriate shaping of other areas. Obviously, the scope of such activities may vary and depend not only on the needs, but also on the legal and actual possibilities of doing spatial planning. Lack of such actions leads to spatial chaos and serious financial losses, some of which are highly noticeable. In the current legal state, however, there is some confusion on the part of the legislator as to how to specifically protect and shape spatial order.

The previous paragraphs highlighted the different approaches to 'spatial planning' in the different CEE countries. Hence, there are countries where 'spatial planning' is not directly defined in the normative framework governing their planning systems. However, there are principles expressed in the laws of some CEE countries that equate spatial planning with 'process', 'instrument', and 'action'. This undoubtedly reflects theoretical perspectives that permeate the dedicated literature. Across these differing approaches, a predicament seems to arise with the conceptual and the implementation levels mentioned a bit earlier: the link between development planning and land use. This is one of the most serious and interesting challenges. Hence, the question becomes how to effectively implement development policy objectives, using each country's normative framework.

The objectives of spatial planning are also defined in different ways. They are often linked to sustainable development, even though such an approach generally remains undefined from a legal perspective. There is also a tendency to link spatial planning with property management. The objectives for a given site are defined in various ways: efficient use of the land, conditions for spatial development, spatial structure and territorial cohesion, or spatial order. This is a fundamental issue: what should spatial planning provide for an area in the first place? From the numerous answers provided in the literature, it appears that several objectives are here at stake. Spatial order, which is linked to the functional and architectural dimension, may be considered a priority. However, it must be linked to the social, economic, environmental, and cultural perspectives. This is why balancing conflicting interests receives such careful attention in legal parlance, with the public interest given explicit priority (note, however, that the public interest is also open to different interpretations, such as the discussion by Moroni 2017).

Another issue is the range of additional topics related to spatial planning. Environmental and nature conservation clearly play a key role, both from a theoretical and a national perspective. Provisions related to the protection of cultural heritage can be seen in a similar way. The same applies to normative frameworks where spatial planning is a key response in addressing the challenges of climate change. Moreover, there is no doubt about the necessity to translate the aims of development policy into spatial planning objectives (e.g., the "territorial disparities" postulate).

However, considering the diversity of approaches across countries notwithstanding, there is also some common ground:

- Spatial planning must accommodate numerous, mutually differentiated objectives. This distinction is both thematic (urban, legal, and environmental perspectives, among others) and technical (development policy and specific land use guidelines). In part, this task is related to the demand to reconcile private and public interests, but it is much broader in scope.
- One important element is the separation of the conceptual level and the implementation level in spatial planning. Linking them in any system is one of the major challenges normative frameworks face.
- The range of thematic issues related to spatial planning is always open to debate. However, there is no doubt about the important role of environmental and nature protection, the protection of cultural heritage, and the response to climate change.

1.3 Spatial Plans

Spatial plans are even more diverse than the meanings of spatial planning. Nevertheless, we can tentatively assume that on the local level they are plans that (1) specify the intended use of land, and (2) more detailed land use parameters (e.g., building height or building intensity) (Nowak et al. 2022a). The specific national solutions will appear in the following chapters. It is, however, helpful to introduce here the different types of plans (spatial plans, land use plans, zoning plans), as well as their national traits. We have therefore tried to identify the dominant type of plan for each country, without losing sight of other plan types, their layouts, and their relationship to development planning. For the purposes of comparison, we focus primarily on spatial plans at the local level. It is worth noting at this point that in part of the systems local spatial plans are divided into general plans and detailed plans. In such approaches, both acts are legally binding, with general plans defining zones and selected land use guidelines. From the perspective of some of the countries surveyed, these solutions are seen as positive. However, there are countries for which general plans are spatial and inadequate to meet current needs. The approach to 'spatial development plans' also varies. Development strategies cannot be equated with this concept. In some national systems, however, the 'spatial development plan' is a separate spatially oriented strategic document. The concept of 'land use guidelines' can also vary. The extent to which these 'guidelines' are legally binding depends on the legal formula of the specific spatial plan in the country concerned.

In Lithuania, spatial plans are territorial planning documents. This includes both comprehensive plans and detailed plans. They can be drafted at different levels. Among other things, they designate individual zones and development parameters. The term 'land use', on the other hand, refers more to land readjustment projects, such as the creation and rearrangement of plots. There is also no independent concept of zoning plans. Zoning is included in spatial plans. *Hence, in Lithuania, the dominant concept is the spatial plan, appearing at different geographic scales and focused on defining zones and development parameters.*

In Latvia, there are three types of spatial development planning documents—spatial plans developed only at the local-level and long-term sustainable development strategies including spatial development perspectives and mid-term development programmes elaborated at all three planning levels: national, regional, and local. Strategies with spatial development perspectives and programmes define appropriate to the level sustainable and spatial development priorities, guidelines and sets of implementation measures. Spatial plans determine land use—functional zoning, land use and building requirements and specify public infrastructure. The concepts of land use plans and zoning plans are not used in Latvia. Zoning of the territory itself, also in the graphic sphere (maps), is included in the local government spatial plans. *In Latvia, spatial development planning documents (strategies and programmes) at all three levels are crucial. Spatial plans, on the other hand, appear at the local level (this concept includes zoning plan).*

In Estonia, a 'spatial plan' designates an overall spatial solution prepared for a certain area, which usually determines the land use and development conditions for that area. A spatial plan consists of an explanatory memorandum and technical drawings, which are drawn up based on the planning work proper. There are five types of spatial plans: national, county, local comprehensive, local detailed, and special. This is a strongly hierarchical system. There are no separate spatial development plans, but the national spatial plan performs part of their functions. It defines the policy and trends for sustainable and balanced spatial development of the entire country. The aim of the plan is to provide a spatial basis, based on environmental characteristics, for the formation of settlements, mobility, the country's technical infrastructure, and regional development. There are no separate land use plans, with land use objectives and development parameters primarily defined within spatial plans at the local level. There are also no separate zoning plans in Estonia. *In Estonia, the key concept is the spatial plan, which includes most of the other types of plans. There is considerable diversity in the categories of spatial plans.*

In Hungary, the spatial planning system is covered by two types of acts: regional development strategies and land use framework plans. There are three levels: national, county, and local. On each level both documents are present. In this perspective, 'land use' includes the establishment of land use principles, the identification of resources, the determination of landscape pressures and its carrying capacity, proposals for the appropriate use of territorial resources, and the definition of a spatial, technical, and physical system of development concepts and programmes.

Starting with 2022, a local general plan in two parts is adopted at the local level: the local development plan and the land use plan. The land use plan aims to define the spatial structure of technical and infrastructural systems, the long-term spatial structure of the area, land use principles, as well as a study of environmental, social, and economic impacts. The land use plan is divided into a structure plan (structure of the settlement network, the spatial structure of land use) and a regulation plan (zoning of the area with specific regulations and restrictions). There is a serious hierarchical dependence of plans. There is no separate concept of a zoning plan in Hungary—it can be understood as a graphic element of the land use plan. *In this system, the broadest term is the spatial plan. The land use plan (a key concept in the system) is an element of this plan, but has a broad scope, strongly linked to development policy.*

The Romanian planning system uses the term 'spatial plans', albeit somewhat implicitly. The terms 'spatial development plan' and 'land use plan' do not exist independently. Like the Hungarian model, the Romanian planning system also works with three levels: the national, county, and local. Each of them must draft:

- Development strategies,
- Spatial plans (at the county level the county spatial plan and at the urban level the general urban plan),
- Investment and operational programmes.

Spatial plans at the regional and national level are not universally binding, in the sense that their provisions are legally binding to the public administration only (internally binding). In contrast, urban plans are universally binding and come in two forms: the general urban plan and the zonal urban plan. They are regulatory plans. The general urban plan contains long-term, medium-term, and short-term regulations. The latter include designation of the built-up area, land use guidelines, development parameters, but also guidelines related to the transport network or protected areas. The medium-term and long-term provisions are geared towards identifying key challenges and problems. Zonal urban plans define the organisation of the road network, the architectural and urban layout of the intervention area, its land use pattern, and the legal status of plots, as well as urban planning restrictions and prohibitions. There is a link between spatial plans, urban plans, and development planning. An example is the requirement to include an urban strategy in general urban plans. *Urban plans play a key role in this system, incorporating zoning, land use rules, and other elements.*

In the Czech Republic, the term used is 'spatial plans'. There are two types of spatial plans: non-statutory planning materials (directional, non-binding) and binding plans (at the national, regional, and local levels). The terms 'spatial development plans', 'land use plans', and 'zoning plans' are included in these formulations. At the local level, there are two types of plans: municipal spatial plans and regulatory plans. The municipal spatial plan defines the development concept for the municipality, protects its values, and designates development areas. It also defines permissible, restricted, and prohibited uses for individual areas. Regulatory plans, on the other hand, lay down detailed conditions for the designation and use of land, including the location and siting of buildings (thus replacing planning permission in a selected built-up area). At the same time, spatial plans at higher levels set out broader development principles and rules. *The dominant concept of spatial plans includes other types of plans, which are not easily separated.*

In Slovakia, territorial plans or land use plans are legally binding documents. Strategic plans are non-binding, they are predominantly used for the implementation of short-term activities and for claiming financial support. At the local level, there are two types of spatial plans: general plans and specific plans. It is not possible to obtain planning permission for most developments if there is no spatial plan approved in that area. *In Slovakia, strategic planning is very weak. Consequently, there is a weak link between spatial and strategic planning.*

In Bulgaria, the concepts of land use plans and zoning plans are also included in the concept of spatial plans. Spatial plans are prepared at the local level and mostly focus on the regulation of public spaces and building parameters. There are two types: general plans and specific plans. They are not linked in a binding way to other programmes. In addition, there are also spatial development plans, which are strategic documents with a spatial component. They include national and regional spatial development strategies. At the municipal level, development plans are also prepared, albeit with separate strategic objectives. Their implementation is carried out through a list of projects and the associated financial instrumentation. However, there is no link between development plans and spatial plans. *In the case of Bulgaria, the concept of spatial plans prevails. However, there is no link to development policy.*

In Poland, the term 'spatial plan' is also dominant. At the local level, there is one type of spatial plan, which defines both the purpose of a given area and the detailed principles for its development (as well as the principles of protection of the environment, nature, or cultural heritage). It is a binding document which incorporates the terms 'land use plan' and 'zoning plan'. A spatial development plan may be regarded as a study in spatial development conditions and directions—non-binding concepts for the spatial development of municipalities (it should be distinguished from municipal development strategies). At the regional level, there is a spatial development plan for the voivodship, which is not universally binding. At the national level, there is no spatial plan. The legislator is trying to broaden the relationship between spatial plans and development strategies, albeit with moderate success. *In Poland, there is only the concept of spatial plan. There is a very weak link to development policy.*

Based on the previous overview, the 'spatial plan' emerges as the dominant concept across CEE countries. This confirms the validity of the initial proposal for understanding the term. The idea that, at the local level, the 'spatial plan' is a document defining the zoning/purpose of a given area and the principles for its development seems very clear across each country. This interpretation seems accurate even if 'land use plans' are also present in the system. The main differences across CEE countries in approaching the 'spatial plan' are the following:

- The scope of application at different levels: There are systems where the spatial plan only occurs at the local level, and there are systems where it occurs at different levels.
- A differentiated binding force: In every system, at least one spatial plan is a universally binding legal act. However, this is not always the case at every level (national/regional).
- Numbers at a given level and scope: There are systems where the spatial plan combines development parameters with a development policy concept for the area. There are also systems where the spatial plan contains only specific zoning provisions and development parameters. Furthermore, spatial plans can be divided into two types at a given level. The basic division distinguishes between general plans and specific plans. Theoretically, the former define the zones, and the latter the principles of land development. In practice, however, individual provisions within these plans differ across countries (e.g., in general plans there are provisions concerning land use). Because of these differences, even a working division of spatial plans into zoning plans and land use plans would not be entirely comprehensible.
- Approaches to the relationship between spatial planning and development policy also vary. When translated into specific documents, CEE countries have different approaches. There are systems where development policy issues are part of spatial plans, while in other systems there are spatial development plans (alongside spatial plans), i.e., development documents which focus directly on the spatial context. In still other cases, the spatial context is included in development strategies (at different levels). In the latter two cases, a particular challenge is to ensure the link between development objectives, on the one hand, and zoning and land use principles, on the other.

To sum up, we can conclude that there are no great differences between the countries in terms of terminology or general approach. The regulatory role of spatial plans at the local level and their general scope are similar. Differences relate to the detailed scope of individual plans, the number of plans at supra-local levels, or the relationship between spatial planning and development policy (which requires a separate, more extensive commentary).

Based on these considerations, we suggest that, despite the numerous analogies, a more in-depth comparison of the spatial planning systems in CEE countries should begin with specific issues rather than specific documents. Simply put, it is more convenient and transparent to describe the way in which land is zoned or the guarantee of development parameters than to compare all (differentiated) documents. These documents should appear in the analysis as a response to specific issues and challenges. The next step then is to compare their detail and their scope, and especially the actual problems occurring in their application.

1.4 Spatial Planning and Development Policy—Selected Dilemmas

As indicated above, the relationship between spatial planning and development policy can be defined in different ways across planning systems. The authors focus on comparisons between CEE systems, while noting that broader classifications can be constructed (OECD 2017).

A discussion on the relationship between development policy and the detail of spatial planning regulations is present in the literature (Savini and Aalbers 2016; Moroni et al. 2020; Buitelaar 2012). There is no doubt that overly detailed rules block development (and hinder the integration of development policies). This is related both to the problematic nature of amending such regulations (in response to emerging challenges and needs) and to problems in interpreting these regulations (Needham et al. 2019; Nowak et al. 2022b). Problems with interpretation occur at each level of detail within such provisions. The intention of the legislator to specify all planning details may lead to additional, unforeseen problems, and increase uncertainty in planning. These observations apply both to spatial planning regulations at the statutory level and to spatial plans which are legally binding acts (and, depending on the spatial planning system, any equivalent of binding spatial plans).

Problems also arise because spatial conflicts prompt individual space users to use spatial planning law in ways that benefit themselves (and their objectives). This stand contributes to a widening of the possible variants of legal interpretation, and consequently to the resulting chaos. The problem is further compounded by changing development needs that are difficult to correlate with the current wording of the legislation. In addition, further challenges, especially those concerning climate protection or the integration of health care with spatial policies, require a very flexible spatial response. Against this background, a large part of detailed regulations, as well as the procedural difficulty of their alteration, become a serious barrier.

Hence, according to the authors of this monograph, there are several barriers, or rather discussion areas, concerning the relationship between spatial planning and development policy (planning). Problems arise not only from regulatory details. There are also difficulties in implementing integrated development planning, its shifting meaning across different disciplines and countries notwithstanding. Basically, this is primarily a matter of determining how the objectives of development policy and spatial development policy are translated into provisions within spatial plans (and consequently, into actual implementation). It is worth pointing out that the aims of development policy in this approach should be understood broadly and be related to, inter alia, climate change, public health, demographic trends, and social problems, as well as environmental protection and the preservation of cultural heritage.

Another aspect is the relationship of the local level to the higher levels. Normally, development policy objectives should be more broadly defined in documents at the regional or national level. Again, depending on the system, these connections are shaped in different ways. However, two types of linkage can be distinguished:

- Development and spatial development (not always inclusive) policy guidelines appear in higher-level acts (development documents or spatial plans, or more or less integrated development and spatial development documents depending on the country's case), which are then translated into local acts.
- An ad hoc interference by central authorities, independent of the provisions contained in planning documents.

When integration of development policy and spatial planning is key, this second type of interference is generally a threat, because the interference is usually aimed at ensuring the possibility of implementing a specific investment or a specific type of investment. Hence, it undermines the logic of the planning system. An example can be 'special acts' (legal acts bypassing the entire spatial order) or extraordinary administrative decisions. This type of interference contributes to blocking development objectives.

Turning now to content of spatial plans themselves, obsolescence may also become a significant threat. It is usually a situation where plans do not consider current development challenges (e.g., because they were adopted many years earlier and not updated). There may also be situations in which spatial plans do not specify development goals. Yet another issue might arise with procedures related to public procurement, both in terms of preparing plans and implementing specific investments, which significantly prolong the perspective of implementing development policy.

When it comes to the dilemma concerning the relationship between the level of regulatory detail and its impact on development policy, we must agree that regulations which are too detailed usually block development. Note, however, that insufficiently detailed regulations also block development, especially in CEE countries. The challenge in each country is therefore to define the regulatory level of detail in such a way, that one is then able to implement development policy. An overly general legal framework, even if linked to the concept of flexibility, determines opportunism, because it implies adjusting to the subjective expectations of investors, disregarding considerations of public interest or development policy. It is also important to stress here that the relationship between spatial planning and development policy may be negatively influenced by the claimant groups of space users. A target postulate to limit such tendencies may be a deeper cooperation of the public and private sectors.

There is no doubt that an adequate relationship between spatial planning and development policy will not be ensured by statutory guidelines alone. Even the requirement to take certain objectives and assumptions within development policy into account when devising spatial planning instruments does not guarantee the achievement of this goal. Even such guidelines may either lead to another case of chaos in the interpretation of regulations or to ignoring the guidelines altogether at the implementation stage.

1.5 Classifications of Spatial Planning Systems—A Contribution to the Discussion so Far

Based on the previous sections, it is now possible we can now expand on the classification of spatial planning systems. As we have seen, there are both discrepancies and analogies in the understanding and application of spatial planning, spatial planning objectives, spatial plans, and the relationship between spatial planning and development policy. Now we can concentrate on relating these issues to the comparative study of spatial planning problems.

There is no doubt that comparing spatial planning systems across different countries is a very difficult task. The first barrier is comprised of diverse legal provisions, which are difficult to compare comprehensively, due to differing legal systems (and different ways of interpreting provisions) in each country. In addition, CEE countries have different planning traditions, since they belong to diverse, but not always internally consistent, planning families. For example, Latvia and Estonia have more Nordic planning culture influence. Hence, great care must be taken when making comparisons. Seemingly similar institutions may be interpreted quite differently, due to distinct cultural traditions or different languages. A second barrier arises due to different planning practices, which usually have a greater influence on spatial planning than the law itself. Related to this is the level of planning culture. An interesting analysis by de Vries (2015) shows that even a very high similarity of laws in different countries does not determine the identity of their planning systems. And yet, a high degree of similarity in spatial planning systems is also very rare. Therefore, one must be aware that comparing both general and specific solutions in different countries will always be fraught with great risk and many understatements. This does not mean, however, that comparative analyses should not be carried out. On the contrary, they are very much needed.

In a European context, such an exercise is subject to diverse demands, with the most important being the Europeanisation of spatial planning. This should be understood primarily as the way in which sectoral EU policies such as competition, economic development, agriculture, nature conservation, environmental protection, or air quality influence national planning systems (Evers and Tennekes 2016; Faludi 2018). Still, this does not imply a homogenisation of national spatial arrangements (Cotella 2020; Stead 2013). Mironowicz and Geppert (2017) mention among the key (European-determined) directions influencing national spatial policies, the reduction of unnecessary energy consumption in mobility, land consumption, the protection of the European cultural heritage of cities and villages, and the new specificity of spatial management in increasingly fragmented and polarised regions. Similar insights are provided by Auziņš (2018), pointing out that European spatial planning systems are increasingly determined by the economic crisis, ongoing reforms, and the challenges of globalisation (similarly, Silvan 2017). Szlachta and Nowak (2021) express hope that the discussion of the Territorial Agenda 2030 and national reconstruction plans may become an important perspective to open a wider debate.

Obviously, spatial planning systems in different countries follow different trends. A distinction is made, for example, between more flexible (UK) and more restrictive approaches to planning content (Moroni et al. 2020; Buitelaar 2012; Muñoz Gielen and Tasan-Kok 2010). The influence of national governments on local spatial planning (Zimmermann and Getimis 2017), and the extent of public participation (Tewdwr-Jones 2017) also tends to vary. A separate issue is the extent to which planning norms are implemented at the investment stage, and the link between spatial planning and other sectors (Nadin et al. 2018).

Authors have therefore proposed a comparison using differentiated criteria (Newman and Thornley 1996; European Commission 1999a, b; Alterman 2010). The delineation of planning families must consider the mutual variations of numerous arrangements within these families. In one of the most recent in-depth analyses, Reimer et al. (2014) have proposed issues such as challenges to the spatial planning system at the macro-level, and the extent of rigidity/flexibility of the system and planning practice at the micro level. Clearly, each of these aspects needs to be clarified. This is illustrated by the varying methodological dilemmas within these classifications. The system types proposed by Newman and Thornley (1996) or the European Commission (1999a, b) can be important reference points, but currently require additions (Tosics 2010; Silva and Acheampong 2015).

It is also important to bear in mind the broad scope of the term 'spatial planning' and its institutional and sectoral linkages. This is illustrated by the methodology adopted by Kovacs et al. (2013) in which the relationship of spatial planning with nature and landscape conservation was verified in several countries. Alterman (2010, 2011) points out that some of the differences between the different systems cannot be put into fully coherent categories. Some planning solutions in individual countries are grossly uncoordinated with other solutions. It is therefore necessary to balance the rationale between showing a reasonably coherent comparison and 'flattening' the differences.

Nadin et al. (2018) point to the problematic nature of broader comparisons (going beyond individual descriptions of national mechanisms). According to these authors, a major barrier is the comparison of national circumstances and terminology (related to the way individual concepts are understood). One should be aware that differences in legal systems do not explain real similarities in planning practices (Buitelaar and Sorel 2010). Getimis (2012) also points to problems of terminology (the adequacy of the concepts used) and suggests that the key in comparisons is to identify trends rather than snapshots of systems. He also draws attention to classifications of spatial planning instruments, distinguishing between visionary, strategic, framework, and regulatory instruments.

Against this rather diverse background, we believe that when classifying spatial planning systems, special attention should be paid to the following issues:

- The degree of centralisation or decentralisation in each system. Due to the diversity of local government systems, however, this is more of a descriptive criterion, requiring not only an indication of the relationship between public authorities at different levels, but also a more in-depth analysis of the extent of interference of the higher level in the lower level.
- The degree of linkage between acts and documents at different levels, including linkages between spatial planning documents and development policy documents. Again, a broader characterisation is usually required. Nonetheless, it is possible to classify the relationships between individual acts.
- The different levels of planning that exist in any spatial planning system. Firstly, they tend to vary in terms of territorial units. The scope of competencies at each level is also different. A synthetic comparison of planning levels is very difficult. From the perspective of spatial planning objectives, they can be categorised by indicating how issues such as zoning and land use or spatial structures are addressed at different geographical scales.
- Defining the links between spatial planning and development policies. In a synthetic dimension, this should include issues of relationships between different types of planning documents or of integration in one document.
- Development of the spatial planning documents—by experts or by public institutions and their planners.
- Public participation in spatial planning and spatial planning process. This requires deeper, broader analyses of each system. In other words, it is possible to verify to what extent a given system assumes the cooperation and involvement of stakeholders.
- Effectiveness in spatial planning implementation monitoring and control. This is an important element, which also requires in-depth comparative studies.
- Consolidation of spatial planning legislation. This involves determining whether spatial planning issues are covered centrally in a single piece of legislation or in more acts.
- Litigiousness in the spatial planning system. The number of spatial conflicts should also be considered when classifying systems. However, this is also more of a descriptive criterion, which is difficult to put in a concise form.
- The role of graphic representation in spatial planning, especially at the local level.

1.6 Conclusions

The analyses carried out within this chapter have shown possible variations and inconsistencies when comparing CEE planning systems, which should always be borne in mind when conducting comparative research. Nevertheless, it is also worth noting the similarities between these systems. When studying the concept of 'spatial planning', it is important to tackle the terminological confusion and the variability of spatial planning objectives head-on. The good news for future studies is that, despite some discrepancies, a common understanding of the term 'spatial plan' is possible across CEE countries, especially at the local level. The dilemmas concerning the relationship between spatial planning and development policy are similar, even though they look different from the legal perspective in different countries. The guidelines defined by these authors on the realistic scope for comparisons provide a reference point for the following chapters. In the next chapter, the characteristics of different spatial planning systems will appear in a more detailed (tabular) way. In addition, issues requiring more detailed, analyses will receive proper consideration.

References

Abercrombie P (1933) Town & country planning. Butterworth, London

Act of 14 March (2006) Act on town and country planning and building code (Building Act). Parliament Czech Republic, 183/2006 Coll. https://www.mmr.cz/getmedia/9a941cf5-268b-4243-9880-d1b169fb33d6/SZ_angl.pdf. Accessed 2 Sept 2022

Act of 27 March (2003) Act on spatial planning and development. J Laws 741. https://isap.sejm.gov.pl/isap.nsf/download.xsp/WDU20030800717/U/D20030717Lj.pdf. Accessed 10 July 2022

Act of 27 April (1976) Act on land-use planning and building order (the Building Act), 50/1976 Coll. https://www.fao.org/faolex/results/details/en/c/LEX-FAOC183199/. Accessed 3 July 2022

Act XXI (1996) Spatial development and planning. https://net.jogtar.hu/jogszabaly?docid=99600021.tv. Accessed 3 July 2022

Adams D (1994) Urban planning and the development process. Routledge, London

Albrechts L (2015) Ingredients for a more radical strategic spatial planning. Environ Plan B Plan Des 42(3):510–525. https://doi.org/10.1068/b130104p

Alterman R (2010) Takings international. A comparative perspective on land use regulations and compensation rights. American Bar Association, Chicago

Alterman R (2011) Guest editorial: comparative research at the frontier of Planning Law. Int J Law Built Environ 3(2):100–112. https://doi.org/10.1108/17561451111148220

Auziņš A (2018) Key trends and aspects influencing changes into spatial planning systems and practices in Europe. In: Proceedings of the international scientific conference, Latvia University of Life Sciences and Technologies, vol 48, pp 26–35. https://doi.org/10.22616/ESRD.2018.065

Buitelaar E (2012) The fraught relationship between planning and regulation. Land use plans and the conflicts in dealing with uncertainty. In: Needham B, Hartmann T (eds) Planning by law and property rights reconsidered. Routledge, New York, pp 207–218

Buitelaar E, Sorel N (2010) Between the rule of law and the quest for control: legal certainty in the Dutch planning system. Land Use Policy 27(3):983–989. https://doi.org/10.1016/j.landusepol.2010.01.002

Campesino Fernández AJ (2020) Ordenación territorial sobre planificación urbanística: la superposición lógica. Extremadura. Práctica urbanística: Revista mensual de urbanismo. Wolters Kluwer, Cáceres

Cotella G (2020) How Europe hits home? The impact of European Union policies on territorial governance and spatial planning. Géocarrefour 94(3). https://doi.org/10.4000/geocarrefour.15648

Cullingworth B, Nadin V (2006) Town and country planning in the UK. Routledge, London. https://doi.org/10.4324/9780203004258

de Vries J (2015) Planning and culture unfolded: the cases of Flanders and the Netherlands. Eur Plan Stud 23:112148–112164. https://doi.org/10.1080/09654313.2015.1018406

Dühr S, Colomb C, Nadin V (2010) European spatial planning and territorial cooperation. Routledge, London. https://doi.org/10.4324/9780203895290

European Commission (1999a) The EU compendium of spatial planning systems and policies. European Commission, Directorate-General for Regional and Urban Policy, Luxembourg
European Commission (1999b) European spatial development perspective: towards balanced and sustainable development of the territory of the European Union. Informal Council of Ministers Responsible for Spatial Planning, Potsdam
Evers D, Tennekes J (2016) Europe exposed: mapping the impacts of EU policies on spatial planning in the Netherlands. Eur Plan Stud 24(10):1747–1765. https://doi.org/10.1080/09654313.2016.1183593
Faludi A (2018) The poverty of territorialism: a neo-medieval view of Europe and European planning. Edward Elgar Publishing, Cheltenham
Getimis P (2012) Comparing spatial planning systems and planning cultures in Europe. The need for a multi-scalar approach. Plan Pract Res 27(1):25–40. https://doi.org/10.1080/02697459.2012.659520
Healey P (1997) The revival of strategic spatial planning in Europe. In: Healey P, Khakee A, Motte A, Needham B (eds) Making strategic spatial plans. Innovation in Europe, 1st edn. UCL Press, London, pp 3–19
Healey P (2004) The treatment of space and place in the new strategic spatial planning in Europe. Int J Urban Reg Res 28:45–67. https://doi.org/10.1111/j.0309-1317.2004.00502.x
Healey P, Khakee A, Motte A et al (1997) Making strategic spatial plans: innovation in Europe. Routledge, London and New York
Hersperger AM, Grădinaru S, Oliveira E et al (2018a) Understanding strategic spatial planning to effectively guide development of urban regions. Cities 94:96–105. https://doi.org/10.1016/j.cities.2019.05.032
Hersperger AM, Oliveira E, Pagliarin S et al (2018b) Urban land-use change: the role of strategic spatial planning. Glob Environ Change 51:32–42. https://doi.org/10.1016/j.gloenvcha.2018.05.001
Hutter G, Wiechmann T (2021) Time, temporality, and planning—comments on the state of art in strategic spatial planning research. Plan Theory Pract 23(1):157–164. https://doi.org/10.1080/14649357.2021.2008172
Jürgenson E, Auziņš A, Burinskienė M (2017) Land value capture to promote local development in Baltics: a comparative study of Estonia, Latvia and Lithuania. In: Proceedings of the 10th international conference 'environmental engineering', Vilnius, 27–28 Apr 2017
Keeble L (1952) Principles and practice of town and country planning. Estates Gazette, London
Koresawa A, Konvitz J (2001) Towards a new role for spatial planning. OECD Publishing, Paris. https://doi.org/10.1787/9789264189928-en
Kovacs K, Polasky S, Nelson E et al (2013) Evaluating the return in ecosystem services from investment in public land acquisitions. PLoS ONE 8(6):e62202. https://doi.org/10.1371/journal.pone.0062202
Ladzianska Z, Ondrejička V, Ondrejičkova S et al (2019) The impact of participatory planning approach on the quality urban design of former riverbank brownfield sites. In: IOP conference series: materials science and engineering, vol 603. IOP Publishing, Bristol, 022003
López Trigal L (eds) (2015) Diccionario de Geografía Aplicada y Profesional: Terminología de Análisis, Planificación y Gestión del Territorio. Universidad de León, Leon
Luukkonen J (2011) Europeanization of spatial planning. Exploring the spatialities of European integration. Dissertation, Nordia Geographical Publications
Maier K (2014) Changing planning in the Czech Republic. In: Reimer M, Getimis P, Blotevogel H (eds) Spatial planning systems and practices in Europe. Routledge, London, pp 235–255
Mironowicz I, Geppert A (2017) Conversation with the Fox: challenges in spatial development and planning in Europe. DisP 53(2):50–51. https://doi.org/10.1080/02513625.2017.1340625
Moroni S (2017) The public interest. In: Gunder M, Madanipour A, Watson V (eds) The Routledge handbook of planning theory, 1st edn. Routledge, New York. https://doi.org/10.4324/9781315696072

Moroni S, Buitelaar E, Sorel N et al (2020) Simple planning rules for complex urban problems: toward legal certainty for spatial flexibility. J Plan Educ Res 40(3):320–331. https://doi.org/10.1177/0739456X18774122

Muñoz Gielen D, Tasan-Kok T (2010) Flexibility in planning and the consequences for public value capturing in UK, Spain and the Netherlands. Eur Plan Stud 18(7):1097–1131. https://doi.org/10.1080/09654311003744191

Nadin V, Fernández-Maldonado A, Zonneveld W et al (2018) COMPASS—comparative analysis of territorial governance and spatial planning systems in Europe applied research 2016–2018 final report. European Commission ESPON, Luxembourg

Needham B, Buitelaar E, Hartmann T (2019) Planning law and economics: the rules we make for using land, 2nd edn. RTPI library series. Routledge, New York

Newman P, Thornley A (1996) Urban planning in Europe. International competition, national systems and planning projects, 1st edn. Routledge, London and New York.

Nowak M, Petrisor A, Mitrea A et al (2022a) The role of spatial plans adopted at the local level in the spatial planning systems of Central and Eastern European countries. Land 11(9):1599. https://doi.org/10.3390/land11091599

Nowak MJ, Śleszyński P, Legutko-Kobus P (2022b) Spatial planning in Poland. Law, property market and planning practice. Springer Briefs in Geography, Cham

OECD (2017) The governance of land use in OECD countries: policy analysis and recommendations. OECD regional development studies. OECD Publishing, Paris. https://doi.org/10.1787/9789264268609-en

Reimer M, Getimis P, Blotevogel H (2014) Spatial planning systems and practices in Europe: a comparative perspective. In: Reimer M, Getimis P, Blotevogel H (eds) Spatial planning systems and practices in Europe. Routledge, London and New York, pp 1–20

Savini F, Aalbers MB (2016) The de-contextualisation of land use planning through financialisation: urban redevelopment in Milan. Eur Urban Reg Stud 23(4):878–894. https://doi.org/10.1177/0969776415585887

Savini F, Salet W (eds) (2016) Planning projects in transition: interventions, regulations and investments. JOVIS Verlag, Berlin

Silva E, Acheampong R (2015) Developing an inventory and typology of land-use planning systems and policy instruments in OECD countries. OECD environment working papers, 94. OECD Publishing, Paris. https://doi.org/10.1787/5jrp6wgxp09s-en

Stead D (2013) The governance of spatial development in Europe: similar challenges, different approaches? In: Conti S, De Vecchis G, Farinelli F et al (eds) IV EUGEO congress. Program and abstract congress book

Szlachta J, Nowak MJ (2021)Territorial agenda 2030 of the European Union and local spatial policies in Poland. Samorząd Terytorialny 12:7–18 (in Polish)

Tasan-Kok T (2008) Changing interpretations of 'Flexibility' in the planning literature: from opportunism to creativity? Int Plan Stud 13(3):183–195. https://doi.org/10.1080/13563470802521382

Tewdwr-Jones M (2012) National planning in the UK. In: Planning for states and nation/states: a transatlantic exploration, Dublin, 15–16 Oct 2012

Tewdwr-Jones M (2017) Health, cities and planning: using universities to achieve place innovation. Perspect Public Health 137(1):31–34. https://doi.org/10.1177/1757913916677524

Tewdwr-Jones M, Allmendinger P (2006) Territory, identity and spatial planning: spatial governance in a fragmented nation. Routledge, London. https://doi.org/10.4324/9780203008003

Tosics I (2010) Book reviews. Urban machinery: inside modern European cities edited by Mikael Hard and Thomas J. Misa. J Urban Aff 32:393–394. https://doi.org/10.1111/j.1467-9906.2010.00513.x

Virtudes A (2015) 'Good' governance principles in spatial planning at local scale. Procedia Eng 161:1710–1714. https://doi.org/10.1016/j.proeng.2016.08.650
Zimmermann K, Getimis P (2017) Rescaling of metropolitan governance and spatial planning in Europe: an introduction to the special issue—Raumforschung und Raumordnung. Spat Res Plan 75(3):203–209. https://doi.org/10.1007/s13147-017-0482-3

Chapter 2
A Brief Survey of Spatial Planning Systems Across CEE Countries

Abstract This chapter provides a tabular overview of the features exhibited by different CEE planning systems. A brief commentary is provided below the table for each country. We felt that presenting the material within 'national' tables, rather than, for example, 'thematic' tables, would make the text more readable for our audience. Specific issues requiring emphasis (and reference to other systems) are highlighted within the comments. The chapter provides a basic (synthesised and comparative) understanding of the key developments in national spatial planning systems. It is noteworthy that the solutions concerning each country are commented on in-depth, following the same pattern. This makes it possible to compare the different institutions. In characterising the spatial planning systems in the countries concerned, attention has been paid to various aspects. A key one seems to be the context concerning spatial planning at the local level and the related instruments. The context of spatial planning at regional and national levels has also been taken into account. Issues related to the role of the courts in spatial planning and environmental protection in spatial planning play a separate role.

Keywords Spatial planning systems · Local planning · Regional planning · National planning

2.1 Introduction

Chapter 1 provides a broader context to enable a comparison of spatial planning systems across CEE countries. This chapter builds the next step: a comprehensive characterisation of the key features of all the systems analysed. As indicated earlier, comparing solutions across countries is a very difficult task. For this reason, too, comparisons have been separated into specific steps that can be analysed independently (in individual chapters).

Hence, this chapter provides a tabular overview of the features exhibited by different CEE planning systems. A brief commentary is provided below the table for each country. We felt that presenting the material within 'national' tables, rather than, for example, 'thematic' tables, would make the text more readable for our

M. J. Nowak et al., *Spatial Planning Systems in Central and Eastern European Countries*, SpringerBriefs in Geography, https://doi.org/10.1007/978-3-031-42722-0_2

audience. Specific issues requiring emphasis (and reference to other systems) are highlighted within the comments. Issues requiring additional analysis will appear in the third chapter, where relevant case studies also appear. In our opinion, this presentation optimally allows for building a broad knowledge base, both for the main traits of individual systems and their variations.

2.2 Historical and Spatial Characteristics of CEE Countries

Bulgaria, the Czech Republic, Estonia, Hungary, Latvia, Lithuania, Poland, Romania, and Slovakia are linked by historical events. Restricting ourselves only to the period after the Second World War, Bulgaria, the Czech Republic, and Slovakia (Czechoslovakia), Poland, Romania, and Hungary followed a similar pattern of sovietisation (Siedziako 2022). During the first post-war years, the communist formations successively consolidated their position within each country. Of course, there were various stages in the period up to 1989, including a fragmentary liberalisation of the system during the late 1960s and the early 1970s. However, this does not change the fact that, for the most part, authorities had autocratic behaviours, which diffused into spatial planning systems. Of course, there were always nuances.

In some countries, positive system traits were also discernible. For example, socialist planning in Bulgaria has had some notable successes in providing public facilities and developing the national territory. The same applies for Romania. The situation was a bit different in Lithuania, Latvia, and Estonia. They were deprived of their own statehood and subsequently entered the Union of Soviet Socialist Republics (USSR), where even stricter governing rules applied.

For all these reasons, in the period from 1944 to 1989, it is difficult to speak of a classic development in spatial planning systems in the countries surveyed (Markowski and Nowak 2022). Planning solutions with a distinct international flavour did exist, and they were often extensive. Nevertheless, in practice, many spatial planning regulations were ignored, and actual planning decisions were tailored to the expectations of authorities. There was also a definite lack of public participation in planning. In most countries, there was also a tendency to strictly link specific planning provisions to strict deadlines for their implementation. These deadlines were generally impossible to meet. Hence, as time went by, planning provisions became ever more fictitious. An additional problem was that some of them were perceived by the public as an element of communist repression. Nevertheless, interesting discussions on spatial planning issues were held in academia in many countries.

After 1989, CEE countries faced the challenge of reforming spatial planning systems (Kowalewski and Nowak 2018), a process fraught with further difficulties. In some cases, spatial planning was seen as a relic of the communist era. As a kind of rebirth, attempts were made to promote guidelines encouraging the widest possible freedom of construction. For all countries, integration into the European Union was a hugely important historical development. Obviously, the European Union does not impose a specific model for spatial planning. Nonetheless, integration has enabled cohesion policy, spatial justice, or environmental issues to be much more widely used in national public debates. Such issues have significantly influenced the discussion on spatial planning systems as well. The same applies to Europe's regional policy, environmental legislation, as well as other sectoral policies.

Against this background, Altrock et al. (2016) draw attention to the diversity of the New Member States (NMS) (this applies to a somewhat broader group of countries). In their view, it is difficult to speak of a single coherent planning family. Among the differentiating factors, they mention administrative differences, the role and position of local authorities, the extent of spatial and social diversity, as well as the different characteristics of housing markets. Davoudi (2007) points to economic differences. Stead and Nadin (2010), Cotella and Janin Rivolin (2011) posit that accession to the European Union is generating a process of 'Europeanisation' in spatial planning across NMS. At the same time, they warn that this is not equivalent to a convergence of planning systems or a convergence in planning approaches. In a somewhat similar vein, Dabrowski and Piskorek (2018) emphasise the influence of the European Union on adopting a territorial perspective when discussing economic development. Some authors even saw the accession of the NMS as an opportunity to develop a new approach to spatial planning (Pallagst and Mercier 2007; Cotella 2007). However, these hopes seem only partially fulfilled at the time of writing (2023).

Hence, common features of CEE countries include (Nowak et al. 2022):

- The difficulty in devising responses to market challenges and in identifying solutions to reduce spatial conflicts.
- An emphasis on the rights of property owners in the spatial planning system, which basically amounts to a sort of 'unwinding' after the communist era. In extreme cases, this implies a very far-reaching building of freedom.
- Inadequate responses to intensive urbanisation and, usually, urban sprawl.

Most of these similarities are a consequence of communist times, when compact urban development was always pursued (Albert et al. 2020), and agricultural land was spared (Feranec et al. 2017).

2.3 Spatial Planning in Bulgaria

The current Bulgarian Spatial Planning Act has been in force since 2001. However, the problem remains that spatial planning is still rather dispersed, with planning provisions included in other laws as well. Spatial planning issues are therefore highly fragmented, and therefore somewhat incoherent. In addition, discrepancies between spatial planning and development policy acts are visible as well (Table 2.1).

Table 2.1 A synthetic view of the spatial planning system in Bulgaria

Information on local government units in the country (which ones are there?). How many units are there at the lowest level (usually municipalities)? What is the area of the country and the population of the country?	According to the Constitution, the territory of Bulgaria is divided into districts (*oblasti*) *NUTS 3* and municipalities *LAU 1*. The municipalities are the basis/main administrative-territorial units for local self-government as they are directly elected every 4 years. District governance is chosen by the national government A municipality consists of one or more neighbouring settlements. When a municipality is a bigger city, it is subdivided into smaller regions (raion) that have directly elected mayor, but not a relatively dedicated budget Bulgaria has 265 municipalities (LAU 1) and 28 districts (NUTS 3). The population as of 31 December 2021 is 6,838,937 people. Surface area: 110.994 km^2
Brief description of the local level of public authorities. Brief indication of which authorities participate in spatial planning by drawing up instruments (the local level can be diverse. It can be municipalities, but also counties and cities)	Municipal councils (decisive and controlling body) and municipal mayors (executive power) are elected in the municipalities. Local self-government is expressed in the right and the real possibility of citizens and their elected bodies to independently decide all issues of local importance. This includes spatial planning and spatial development of municipality and the settlements within it (Local Self-Government Act, Art. 17). The municipal council adopts resolutions/decisions on the creation and approval of spatial plans and their amendments for the municipal territory or for parts of it, under the conditions and in accordance with the Spatial Planning Act [2]; it also adopts strategies, forecasts, programmes and plans for the development of the municipality, which also reflect European policies for the development of local communities
Characteristics of legal acts concerning spatial planning	In Bulgaria, the Spatial Planning Act (*Zakon za ustroistvo na teritoriata*) since 31 March 2001 is in force Spatial planning is also partially covered in the discourse of other territorial issues and their legislative frameworks, such as in the Black Sea Coast Act, the Regional Development Act, the Environmental Protection Act, the Forests Act, the Water Act, Roads Act, Railway Transport Act, and a lot of ordinances of a lower order An overall observation is that the spatial policies are quite fragmented between different sectoral regulations—building, environment, agriculture, ownership, road, rail and water infrastructure, etc. This fragmentation continues by adding more Acts when needed by the Europeanisation processes. Hence, this fragmentation and poor integration happens as well for the spatial planning instruments—plans, development plans, strategies, etc. Conflict: Spatial planning in the Spatial Planning Act and the Regional Development Act creates conceptually different frameworks of action and interpretation. In the Regional Development Act, spatial planning is treated as strategic planning, while in the Spatial Planning Act as land use planning Conflict: The Spatial Planning Act treats urban and territorial development as buildings, but does treat poorly linear objects, e.g., railway infrastructure

(continued)

Table 2.1 (continued)

Determination of which acts at the local level define the zones (and whether these acts are binding and there is an obligation to enact them)	In Bulgaria, zoning land use is defined by local general spatial plans—residential, industry, mixed-use, infrastructure, blue and green zones, etc. General spatial plans are not a sufficient basis for investment. The municipal authorities are obliged to adopt them according to the SPA. Often they are adopted without prioritising zones or attached financial instruments that would guarantee their follow-up Detailed plans are necessary to execute the investment project. Detailed plans address specific projected investments—they can be adopted on the scale of a neighbourhood or simply on one or two plots In many municipalities where a general spatial plan is not enacted (despite such a legal obligation), a detailed plan may be enacted (this is the original legal trick) *Municipalities will not have the right to approve detailed spatial plans without an adopted general plan. The requirement is expected to enter into force on 1 January 2023*
Identify which acts at the local level set out detailed guidelines for land use development parameters (and whether these acts are binding and there is an obligation to enact them). Identify what these guidelines are	The general spatial plans organise the basis for development on the territory of the municipalities, of parts of them, or of separate settlements into these areas. These plans determine: the general structure and the predominant land use definition of the territories, the visibility and the requirements of the technical infrastructure and the impact on the environment and the sites of the cultural and historical heritage, with retention in the preparation; implementation of landscape measures and aesthetic shaping; etc., of detailed components of plans. In this way, guidelines for land use are laid down in local spatial plans (general and detailed). General spatial plans determine basic development parameters—density, gross floor area, height and green space coefficient. Detailed spatial plans determine the public space design, the exact urban forms of each building, underground infrastructure, etc. It happens that for one and the same territory the detailed and the general spatial plans give different development parameters which have created conflicts and extreme densities at some particular areas (e.g., new neighbourhoods in Sofia) Both these plans are quite rigid, they are adopted by municipal councils and can be changed with difficult lengthy procedures
Determine whether a graphic/ map is part of the spatial plan at the local level. If yes—is it a binding part	Spatial plans at the local level include a text and a graphic part. Data from topographic maps, cadastres, levelling plans, specialised maps and registers and others in digital and graphic form, as well as other data from the specialised information systems of central and territorial administrations and operating companies, are used for the development of spatial plans At the request of the mayor of the municipality, other departments or interested parties, the creation of specialised maps, registers and information systems with specialised data can be carried out simultaneously with the creation of the cadastral map and cadastral registers The detailed development plan is drawn up on the basis of a cadastral map, approved in accordance with the Law on the Cadastre and Land Registry The graphical parts of both general and detailed spatial plans are binding and are with the judicial power as laws. For example, the line separating a road and the attached sidewalks is determined by the detailed plans and cannot be changed—the road lanes cannot be made narrower and the sidewalks cannot be enlarged without changing the plan

(continued)

Table 2.1 (continued)

Identify which acts at the local level shape spatial development policy (if different acts, what is more or less in each document)	Development policy at the local level is defined by: • Local development plans and local spatial development concepts (with strategic development planning until 2020) • Sectoral plans, programmes, and strategies (mainly strategic planning documents) • Integrated development plans and strategies (strategic spatial development since 2020) The plan for the integrated development of a municipality defines the medium-term goals and priorities for sustainable development of the municipality and its relations with other municipalities in accordance with the integrated territorial development strategy of the planning region (NUTS 2) and the general spatial plan The plan for the integrated development of a municipality provides spatial, temporal, and factual coordination and integration of various policies and planning resources to achieve the defined goals for permanent improvement of the economic, social, and ecological condition of the municipal territory
Regional-level characteristics	The regional level in Bulgaria consists of districts (*oblasti*) (administrative unit; NUTS 3; 28 districts) and planning regions (NUTS 2 level units; not administrative units) Spatial Planning Act> At the regional level, the concepts and schemes for spatial development determine the objectives of the state policy for the organisation of the territory for a certain period The conditions and procedures for assigning, developing, accepting, and implementing spatial development concepts and schemes are determined in the Regional Development Act Regional Dev. Act> Schemes for NUTS 2 and 3 have been cancelled. Integrated Spatial Development Strategies are currently in place for NUTS 2 planning regions The integrated spatial strategy for the development of a level 2 planning region defines the medium-term goals, priorities, and prospects for sustainable integrated regional and local development on the territory of the relevant planning region in accordance with the provisions of the National Concept for Regional and Spatial Development and others sectoral and horizontal policies, as well as its links with other level 2 planning regions in the country and in neighbouring countries in the macro-region The integrated spatial development strategy of the level 2 planning region coordinates the predictions of the sectoral strategies and documents at the regional level in the fields of economic development, health, education, science, social services, transport, water sector, energy, broadband communications, tourism, and the environment and takes into account their regional specificities and territorial dimensions Conflict: uncoordinated changes by the legislature in both acts
Identify which acts at regional level set out zoning and land use guidelines (or related elements)	No act specifies zoning or land use guidelines
Identify which acts at the regional level define development policy in the spatial dimension	The integrated spatial development strategy of the level 2 planning region
Brief overview of the relationship between acts at regional level and acts at local level	The regional integrated development strategies must be taken into account in local integrated development plans Weak relationship

(continued)

Table 2.1 (continued)

Identify (if any) which acts at national level relate to spatial planning matters	In Bulgaria, there are two documents at the national level: The National Concept of Spatial Development (2013–2025) and the National Strategy for Regional Development (2012–2022), with the national concept being the instrument that *aims* to coordinate the territorial development processes and sectoral policies. The national concepts flag general spatial matters. Both documents are governed by the Regional Development Act and both are strategic (not regulatory) documents. Spatial planning issues are concerned in both documents
Identification of specific solutions	Example from Sofia: There is an attempt by the municipal planning office Sofia plan to bind the general spatial plan and the integrated municipal plan by connecting them in a single process of creating a common vision, gathering data analysis from different departments, and of determining common goals and solutions that can be integrated and synchronised as priorities in both plans. This process is still in its very beginning but it can be an example of forcing an integration of the spatial plan and the integrated strategic plan into one single institutional effort Example from general plan of Koprivshtitsa (2014): The authors of the general spatial plan decide to open up the process of creating the plan to the public by organising a wider series of public discussions and working with focus groups. The legal framework proposes limited solutions of spatial actors' participation/involvement when creating the general plan, but the authors opt for a sustainable process that would search for a wider coalition. As a result, the plan proposes that the town of Koprivshtitsa and its future development rest in the current urban limits while at the same time a strong emphasis is put on the conservation of the architectural heritage in the municipality
Identify how environmental and nature conservation sections are included in spatial planning acts at local level	There are no environmental or nature conservation sections in the Spatial planning Act. Nevertheless, the main instrument regulating these aspects of spatial policy is the environmental impact assessment, demanded for all spatial and development plans, strategies as well as for bigger investment and infrastructure projects by the Environmental Protection Act A lot of spatial plans and strategies are unable to pass through a positive EIA thus are subjected to modifications
Problems with spatial plans at local level	The general or detailed spatial plans in Bulgaria often do not have financial instruments attached to them that would secure a prioritisation or execution of their programmatic goals. If new neighbourhoods are planned, the private investments would happen, but there would be no guarantee that there will be a follow-up with the necessary infrastructure—land expropriation, street/water/sewage infrastructure, schools/kindergartens, social infrastructure, etc. Municipalities would have small budgets for expropriation or infrastructure. But the chief architects (responsible for issuing building permits) are not allowed to reject a permit application that is aligned with the law. This puts the investment pressure on single municipal departments (chief architects) but does not resolve the problems of underfinanced the creation and maintenance of social and technical infrastructure that should follow the investment processes Urban planning has been often delegated also to chief architects, which makes their department responsible at the same time for the planning and its execution via permit procedures. Thus, planning rests at the margins of the capabilities of these departments and is often delegated to private planning offices Example 1 from Sofia: The last general spatial plan has set enormous private territories for parks and green zones. The law gives the municipality the possibility to expropriate these terrains in a 10 or 15 years horizon but the budgets had never been set in order to complete these goals. Hence, now some enormous green territories and parks are threatened of being fenced or even constructed

(continued)

Table 2.1 (continued)

Defining the role of the courts in relation to planning acts at local level	Spatial plans do not have established practice of negotiating with the general public or do not create public platforms to negotiate conditions and content transparently with owners and other actors in the process. Thus, too often plans and projects for public infrastructure end up contested in court Judicial practice plays quite big role in taking decisions as the court could take decisions on the changes of Spatial planning Acts or on the content of the Spatial plans
Frequency of changes in the spatial planning system	The Spatial Planning Act has been one of the most changed and amended Act—103 changes and amendments for 21 years which makes almost 5 times per year. The investment pressure has made construction possible a lot easier but as well it had weakened the authorities and their controlling mechanisms

Source Own elaboration

Local government units in Bulgaria vary in size. At the local level, two types of municipal public authorities can be distinguished: municipal councils and municipal mayors. At the local level, municipal authorities enact and apply key spatial planning instruments, while at the same time being responsible for determining the direction of spatial development at the local level. The key instrument at the local level is the general spatial plan, which is a comprehensive plan. It acts as the main zoning ordinance. However, detailed plans are the real basis for the implementation of investments.

In practice, one of the main problems is that municipalities do not adopt general plans and thus do not engage in zoning, despite a formal obligation to do so. They then (deliberately) make use of the possibility of adopting detailed plans for specific investments. These investments then become exempt from the rigours to which they would be subject if they were to come under the purview of general plans. In 2021, for example, around 40% of Bulgarian municipalities have not drafted and adopted general spatial plans, mainly due to low expertise capacity and weak political will.

Both general plans and detailed plans define the parameters for development. However, the parameters included in detailed plans are more elaborate. Procedural issues with the adoption of both categories of plans need to be highlighted [which are?]. Both types of plans contain legally binding graphic sections, with the role of demarcation lines [what are these? Zoning contours?] being particularly important here.

At the local level in Bulgaria, development policy is also pursued. Some policy issues are included in spatial plans, while others are included in dedicated acts, such as sectoral acts or special integrated development plans and strategies. These instruments deserve special attention: they testify for the insufficient integration of development policies and the attempts made to solve this problem.

At the regional level in Bulgaria, spatial planning has a very general, directional character. No zones or development parameters are set at this level. There are also acts shaping integrated development planning. Previously, the Spatial Planning Act (2001) did define spatial planning instruments at a regional scale, which were never drafted (regional spatial development prospects). In the latest changes to the law, reference is made to the Regional Development Act (2008), which does define instruments, but they are strategic in nature (strategic planning). This could have a double reading: an attempt to integrate regional policy with spatial planning, or the gradual replacement of spatial plans and prospects with regional plans and strategies.

At the national level, there are development planning acts and sectoral acts. Note, however, that environmental provisions are missing from the spatial planning legislation.

As a specific solution for Bulgaria, an attempt should be made at combining the provisions of spatial plans and integrated plans in a single act. This is a response to the emerging challenges in CEE countries concerning the deeper integration of development policies. Combining strategic, urban, and legal perspectives is always a challenging task, requiring both conceptual and terminological coherence.

The main problems with spatial planning at the local level in Bulgaria are related to the lack of connections to financial instruments. Consequently, this makes it difficult to align planning concepts (contained in the plans) with financial possibilities, for example provisions related to technical infrastructure. Another issue is the overburdening of architects responsible for urban planning (mainly chief architects) with additional responsibilities, thereby hindering the architectural realisation of planning objectives. This is further compounded by the important role of courts, which resolve complaints against spatial plans and assess the content of spatial plans from a legal perspective. The legislator has made frequent attempts to correct the planning system, resulting in a series of rather frequent amendments of spatial planning regulations.

2.4 Spatial Planning in Czech Republic

The Czech Spatial Planning Act has been in force since 2006. Like the Bulgarian case, the Czech legal system spreads spatial planning issues across different laws. Again, this poses a predicament when interpreting legislation. Zones are singled out in local spatial plans. More detailed issues may be dealt with in separate regulatory plans, which are also adopted at the local level (Table 2.2).

In the Czech Republic, we can distinguish between municipalities and regions. The key instruments for spatial planning work at the municipal level, being the responsibility of the municipalities, where we find a municipal council and a municipal board (i.e., a collegial body, headed by a mayor). The local level is also responsible for defining development directions.

However, municipalities are not required to adopt these two types of plans, i.e., general and detailed plans. The situation is therefore different than in Bulgaria, especially regarding zoning plans. In addition, zoning plans are generally enacted in the Czech Republic. We find binding development parameters for investors in both types of plans. The difference is that guidelines in the general spatial plans are broader, while those in detailed plans are more particular and are linked to technical and construction parameters. Drawings are included in general plans (i.e., in the Czech Republic land use plans) and have a binding character. Here, too, the lines demarcating areas with different uses play a special role.

Table 2.2 A synthetic view of the spatial planning system in Czech Republic

Information on local government units in the country (which ones are there?). How many units are there at the lowest level (usually municipalities)? What is the area of the country and the population of the country?	In the Czech Republic, there are municipalities (6249) and regions (14). Country has a population of 10.70 million. The surface area of the Czech Republic is 78,870 km^2
Brief description of the local level of public authorities. Brief indication of which authorities participate in spatial planning by drawing up instruments (the local level can be diverse. It can be municipalities, but also counties and cities)	Municipalities have a municipal council (adopts resolutions and decrees—the controlling body) and a council board with the mayor—the executive body. The municipal council adopts local spatial policy acts (územní plán a regulační plán). The council board issues decisions on development conditions
Characteristics of legal acts concerning spatial planning	There is Act No. 183/2006 Coll., on Spatial Planning and Building Rules ('Building Act') (includes spatial planning and building order) is in force in the Czech Republic. New Act 283/2021 was approved in 2021 and shall come in force in July 2023. However, issues concerning spatial planning appear in dozens of other acts concerning other themes (e.g., environmental protection, nature protection, and protection of monuments). This often causes interpretation problems
Determination of which acts at the local level define the zones (and whether these acts are binding and there is an obligation to enact them)	Zoning—land use is determined by local land use plans Act No. 183/2006 Sb. (land use means determination of the purpose for which a given area is to be used. It may be, e.g., for housing, services, etc.). Land use plans are binding acts for investors/property owners Regulatory plan defines more detailed regulations of construction on land The municipal authorities are not obliged to adopt neither land use plans nor regulatory plans. It depends on their discretion
Identify which acts at the local level set out detailed guidelines for land use—development parameters (and whether these acts are binding and there is an obligation to enact them). Identify what these guidelines are	Guidelines for land use and other regulations are established in the local land use plans. Land use plans are binding acts for investors/property owners. The land use plans set out the following development parameters: zoning, building density, share of greenery, phasing of development, biologically active area Regulatory plans set out the following development parameters: building height, building line, number of parking spaces, etc.
Determine whether a graphic/map is part of the spatial plan at the local level. If yes—is it a binding part	Land use plans at the local level include a text and a graphic part. The graphic part specifies the intended use of the land and contains the demarcation lines between areas with different uses. It includes also further regulations for dimensions of buildings. The graphic part is also the binding part of the plan
Identify which acts at the local level shape spatial development policy (if different acts, what is more or less in each document)	Development policy on local level is determined by the development principles issued by the region Main tool is the land use plan City may use strategical plan to specify strategic objectives. Strategical plan is not anchored in the legislation The existing planning tools are very weak in the definition of spatial development policy

(continued)

Table 2.2 (continued)

Regional-level characteristics	There are 14 regions in the Czech Republic. Region has the following authorities: the regional parliament (which adopts the resolutions and decrees of the region) and the board of the regional parliament as an executive body (headed by the governor/hejtman of the region)
Identify which acts at regional level set out zoning and land use guidelines (or related elements)	The development principles (zasady uzemniho rozvoje) define the basic elements of the settlement network of the region and their connections as well as translocal public purpose investments. It is a binding act for municipalities for their land use plans. In this sense, it is binding for investors Act No. 183/2006 Coll., on Spatial Planning and Building Rules ('Building Act') specifies content of the Development Principles
Identify which acts at the regional level define development policy in the spatial dimension	The document, the development principles, specifies the objectives in the spatial dimension, the expected results of actions and indicators for their achievement, a model of functional and spatial structure on a regional scale and arrangements and recommendations for the implementation of spatial policy on a regional scale Regions may use strategical plan to specify strategic objectives. Strategical plan is not anchored in the legislation
Brief overview of the relationship between acts at regional level and acts at local level	The development principles on regional level are binding for municipalities when drawing up local land use plans Parts of land use plans which do not correspond fully with development principles of the region are not legally valid
Identify (if any) which acts at national level relate to spatial planning matters	Spatial development policy (Politika územního rozvoje) approved by the government of the state addresses the spatial arrangement and functional use of the territory of the Czech Republic and establishes a framework of social, economic, environmental and cultural requirements on the territorial development of the country and its regions
Identification of specific solutions	There is no other way to plan or permit constructions as per Building Act
Identify how environmental and nature conservation sections are included in spatial planning acts at local level	The principles of environmental and nature protection are defined: • In laws on environmental protection (Environmental Protection Law, Act. No. 114/1992 Coll.) and nature conservation (Nature/Conservation Law, Act No. 17/1991 Coll.) • In Building Act No. 183/2006 Coll. • In laws on agriculture land protection, forest land protections, waste management, air protection act, etc. Spatial Development Policy, Development Principles, land use plans and regulatory plans shall take into account all the contents of those acts There is a system of state authorities which make statement to all spatial planning documentation on regional or municipal level. Disagreements are solved by responsible ministry Plans deal with protection areas (nature, agriculture, forestry, water catchment areas, mineral resources, etc.) and so-called territorial system of ecological stability (shall maintain fauna and flora)
Problems with spatial plans at local level	Main problems of spatial planning in the Czech Republic might be seen in the unbalanced distribution of aims and tasks among the planning tools. Strategical planning is not legally anchored, too many tasks are put on land use plan, and almost no power is dedicated to regulatory plan. This leads to endless proceedings of the land use plans Further, Czech planning is almost helpless in real definition, valuation, and enforcement of public interests in space. There is missing steering culture among authorities protecting individual public interests which leads to contradictory statements Many troubles of the Czech spatial planning results from administrative arrangement of the state. There are about 6250 municipalities, many are extremely small. Coordination of state and regional investments is jeopardised by recalcitrant municipal councils

(continued)

Table 2.2 (continued)

Defining the role of the courts in relation to planning acts at local level	All planning tools may be challenged by property owners, citizens, and civic associations. Judicial review of appeals is sometimes formal not taking into account sense of spatial planning The courts assess not only procedural issues, but also the content of the acts (e.g., they assess whether the scope of land use restrictions is justified). This involves administrative courts to a very large extent in shaping spatial policy
Frequency of changes in the spatial planning system	There have been adopted 26(!) amendments to the Act No. 183/2006 Coll., on Spatial Planning and Building Rules since 2006. Finally, a new building code with principal changes was approved in 2021 and shall come into force in 2023. But the code is severely criticised by municipalities, planners and preservation experts and major changes are announced before its validity date

Source Own elaboration

Development policy at the local level is limited to the implementation of development strategies. This applies mostly to cities. A more prominent regional policy appears at the regional level, where it is possible to distinguish both a planning act, referring to the settlement network and public objectives, which are binding for local spatial planning, and a series of development principles, defining, inter alia, the functional and spatial structure of the territory. However, a problem occurs with the translation of guidelines contained in this act into local spatial planning.

At the national level, there is an act that defines the key national priorities for spatial development. Environmental- and nature-related issues are included in spatial planning and provide a point of reference for spatial solutions. However, there are also linked to the application of additional legislation. Unfortunately, there is an excess of objectives and tasks assigned to individual planning instruments, when viewed from the local. This overabundance introduces a certain amount of conceptual confusion and a lack of realistic feasibility for all objectives. It can be linked, in turn, to the underrepresentation of the public interest in the Czech spatial planning system. Added to this is the important role of the courts, who theoretically focus on formal aspects, but often shape spatial policy in practice, when assessing contested spatial plans. To complicate matters even further, significant changes to spatial planning legislation occur rather too frequently.

2.5 Spatial Planning in Estonia

A recent law on spatial planning has been in force since 2015 when the first spatial planning act was enforced in 1995 that was unified with the Building Act. These two fields were separated in 2002 when two different laws were adopted. Spatial planning issues are also regulated in other acts. Nevertheless, it is crucial to distinguish between spatial planning provisions and the Building Code (Table 2.3).

Table 2.3 A synthetic view of the spatial planning system in Estonia

Information on local government units in the country (which ones are there?). How many units are there at the lowest level (usually municipalities)? What is the area of the country and the population of the country?	In Estonia, there are 15 counties (maakond), which are administrative units without the government There are 79 local governments (kohalik omavalitsus)—15 towns (linn) and 64 rural municipalities (vald) in Estonia. The total area of Estonia is 45,227 km^2, and there are 1.3 million inhabitants Earlier, there used to be county governments, which were level between republic government and local governments. But county governments were dissolved in 1.01.2018 as a part of the state reform. In the planning system, county level still exists and there are county-wide spatial plans
Brief description of the local level of public authorities. Brief indication of which authorities participate in spatial planning by drawing up instruments (the local level can be diverse. It can be municipalities, but also counties and cities)	All local issues are organised by the representative body of a local authority—municipal council (volikogu)—which is elected by the residents of the rural municipality. Municipal council forms the executive body, i.e., municipal administration (valitsus) Local governments have the right to prepare comprehensive plans, local government-designated spatial plans, and detailed plans. Local municipalities are politically and legally responsible for the content of these plans, but their actual preparation is usually outsourced (by developers or land owners) to private consultants
Characteristics of legal acts concerning spatial planning	In Estonia, the (renewed) Planning Act was passed on 28.01.2015. It establishes the principles of planning and the requirements for the planning procedure and for the implementation of spatial plans on all levels. Issues of spatial planning appear also in Building Code (2015) and in An Act to Implement the Building Code and the Planning Act (2015). There are many more acts, which are important in planning process The first Planning and Building Act after restoring independence was adopted in 1995. Separate laws the Planning Act and Building Act were adopted in 2022
Determination of which acts at the local level define the zones (and whether these acts are binding and there is an obligation to enact them)	In Estonia, there are no zoning plans. It is possible to handle the comprehensive and detailed spatial plans as similar to the zoning plans. In the comprehensive plans, the main purpose of how the given area should be used is regulated (e.g., for housing, services, etc.). Detailed plans consist more detailed information about smaller area (measurements for new buildings, etc.)
Identify which acts at the local level set out detailed guidelines for land use—development parameters (and whether these acts are binding and there is an obligation to enact them). Identify what these guidelines are	General guidelines for land use are identified in comprehensive plans. The main purpose of how the given area should be used is regulated. Every local municipality is obligated to prepare a comprehensive plan. More detailed guidelines for land use are written into detailed plan. Local governments can decide if detailed plan is needed in some area or not. If there is detailed spatial plan, then it is binding; otherwise, comprehensive spatial plan is binding
Determine whether a graphic/map is part of the spatial plan at the local level. If yes—is it a binding part	Graphic/map is part of the spatial plan at the local level and it is binding. Planning Act (2015) section 3 subsection 3 enacts that a spatial plan consists of an explanatory memorandum and technical drawings that are created as a result of the planning exercise and that complement each other and constitute an integral whole. The graphic part specifies the intended us of the land and contains the demarcation lines between areas with different uses

(continued)

Table 2.3 (continued)

Identify which acts at the local level shape spatial development policy (if different acts, what is more or less in each document)	Spatial development policy in the local level is shaped by: • County-wide spatial plan • County development strategies • Comprehensive spatial plan • Local development strategies In Estonia, the planning system is hierarchic. County-wide spatial plans are the basis for the preparation of comprehensive plans (Planning Act section 55 subsection 2). County-wide development goals are set in county-wide spatial plans, and comprehensive plans have to follow these goals. It is possible to change the county-wide spatial plan with the comprehensive plan, but these changes must be granted by the Minister in charge of the policy sector (Minister of Economy) County development strategy provides the basis for jointly directing the county development by the local authorities and cooperation partners in the county, planning the jointly made investments, etc. (Local Government Organisation Act section 37^3 subsection 1) According to the Local Government Organisation Act (1993) section 37 subsection 1, all rural municipalities and cities shall have development plans and budget strategies on the basis of which the development of different fields of life is integrated and coordinated
Regional-level characteristics	After the restoration of independence in 1990, there has been established a two-level self-government system in Estonia: (1) parishes, small towns, and towns were on the lower level of self-government; (2) fifteen counties and six republican towns were on the higher level of self-government. In 1994, it went over the one level self-government system in Estonia: parishes, small towns, and towns. The county-level moved under state responsibility In 2017, rapid administrative-territorial reform was performed in Estonia. Because of this reform, the number of local governments has decreased to 79 (there were 213), and the county-level was abolished (Auziņš et al. 2020a, b)
Identify which acts at regional level set out zoning and land use guidelines (or related elements)	Planning Act (2015) establishes the principles of planning and the requirements for the planning procedure and for the implementation of spatial plans, and it regulates all planning levels: national spatial plan, national designated spatial plan, county-wide spatial plan, comprehensive plan, local authority designated spatial plan, detailed spatial plan Planning Act, section 55: The purpose of a county-wide spatial plan is to define the principles and directions of spatial development of the entire county or a part thereof, or of another region. A county-wide spatial plan is prepared primarily in order to express interests that transcend the boundaries of individual local authorities, and in order to balance national and local needs and interests regarding spatial development
Identify which acts at the regional level define development policy in the spatial dimension	According to the Planning Act (2015), there is obligation for the county-wide spatial plan. Local Government Organisation Act (1993) engages the local governments to adopt county development strategy. According to this Act section 6^1 subsection 1 the function of local authorities is to jointly plan the development of the county and to target its implementation. The local-level development plan shall take into account the county development strategy

(continued)

Table 2.3 (continued)

Brief overview of the relationship between acts at regional level and acts at local level	There are strong relation between the spatial plans and strategies and development plans that made for different levels. The county-wide spatial plan must be taken into account when drawing up comprehensive plan on the local (municipal) level. Meanwhile, it is possible make changes from the lower planning level to higher level. It means that changes that are made in the comprehensive plan on local level should express on county-wide spatial plan as well. The local-level development plan shall take into account the county development strategy. County development strategy is worked out jointly by local governments
Identify (if any) which acts at national level relate to spatial planning matters	Planning Act (passed on 28.01.2015) establishes the principles of planning and the requirements for the planning procedure and for the implementation of spatial plans, and it regulates all planning levels: national spatial plan, national designated spatial plan, county-wide spatial plan, comprehensive plan, local authority designated spatial plan, detailed spatial plan In the national level, we have National Spatial Plan Estonia 2030+ and Estonian Maritime Spatial Plan. Estonia's strategic goals are set in the 'Estonia 2035' strategy National Spatial Plan Estonia 2030+ tackles the most general and principal matters of spatial development for the country as a whole. It provides guidelines for county plans and international planning cooperation as well as for spatial development of infrastructure, energy production, etc. (Eesti 2030+) Estonian Maritime Spatial Plan provides guidance and conditions for the next steps in the planning of activities, including those at a local government level. The plan focuses on the principles of spatial development and the activities are not planned in detail (Ministry of Finance 2022) The 'Estonia 2035' strategy sets out five long-term strategic goals that are based on the base principles. These basic principles are set in five categories: people, society, economy, living environment, and governance (Government 2022)
Identification of specific solutions designated spatial plan	In Estonia, there are national designated spatial plan. Planning Act section 27: The purpose of a national designated spatial plan is to erect a construction work which has a significant spatial impact and whose chosen location or whose functioning elicits significant national or international interest. A national designated spatial plan is prepared, above all, to express interests which transcend the boundaries of individual counties in the fields of national defence and security, energy supply, the transport of gas, waste management, or for the expression of such interests in public water bodies and in the exclusive economic zone The local level is local designated spatial plan. Planning Act section 95: A local designated spatial plan is prepared in order to erect a construction work that has a significant spatial impact and whose location has not been determined in the comprehensive plan
Identify how environmental and nature conservation sections are included in spatial planning acts at local level	The Planning Act (2015) determines environmental principles that apply to all planning levels. The General Part of the Environmental Code Act (2011) defines some of the aspects, which must be handled in the comprehensive plans or in the detailed plans. At local level, all the aspects defined in this act, or in the other acts must be taken into account In the process of making the comprehensive plan, it is mandatory to initiate strategic environmental assessment. In some cases, the strategic environmental assessment has to be made in the process of detailed plan also Usually, the environmental and nature conservation sections are in the text and on the plans of comprehensive plan. Also, one of the functions of comprehensive plane is to specify the location and the conditions that ensure the functioning of a green network and to determine the restrictions resulting from such a network (Planning Act § 75 (1) 10)

(continued)

Table 2.3 (continued)

Problems with spatial plans at local level	Details plans are sometimes as separate 'areas', no holistic view for lager area, municipality. It is sometimes connected with the private ownership—one owner land plot means one planning area—it is easier implement, no need to cooperate with other owners. It is connected also the awareness of officers of municipality, and they must keep on the mind the holistic view The implementation of detail plan can be problematical part. There were made changes in Planning Act (2015) ensuring the adopted plan implementation. Planning Act section 128 paragraph 2 point 2 enacts that initiation of the preparation of a detailed spatial plan is forgone above all where it is evident that future implementation of the plan to be initiated is impossible, above all where the authority arranging the preparation of the plan lacks the means to bear the costs of constructing, according to the plan, the roads designated as public roads, together with the related civil engineering works, vegetation and street lighting, or the costs of constructing, according to the plan, the technical infrastructure that serves public interests, and the party interested in the preparation of the plan refuses to bear such costs. It gives the local government possibility to decline the creation of detailed spatial plan The issue of public value capture that can help to implement development is not well elaborated in Estonia. The tool that is used at the state level is land tax; however, it has not supported development classical way and this counting as tool of pubic value capture is even discussable. Every municipality has developed their own tools, and they make the contracts with developers (Taimsaare and Jürgenson 2022)
Defining the role of the courts in relation to planning acts at local level	The Estonian court system consists of four county courts, two administrative courts, two circuit courts, and the Supreme Court. County courts and administrative courts are the courts of first instance; circuit courts are courts of appeal and the Supreme Court in Tartu is the court of cassation and also the court of constitutional review Administrative courts hear administrative matters as courts of first instance County courts as courts of the first instance hear all civil, criminal, and misdemeanour matters It depends on what kind of issue there is, if it is administrative issue, it goes to the administrative court. If the problem is between the private parties, it is civil case, then this issue is going to the county court Spatial plans can be appealed in courts. Anyone who finds that the decision by which a spatial plan is adopted is contrary to public interest or infringes their rights or interferes with their freedoms has a right to contest the decision in court within 30 days following the day on which they became or should have become aware of the adoption of the plan (Planning Act section 54; section 94; section 141) If the court judgement finds that adopted plan is contrary to public interest or infringes the claimant's freedoms and the modifications in the spatial plan are necessary, then the corresponding modifications in the plan must be made by the authority that arranged the preparation of the plan
Frequency of changes in the spatial planning system	The first Planning and Building Act after restoring independence was adopted in 1995. Separate laws, the Planning Act and Building Act, were adopted 2022. Renewed Planning Act passed on 28.01.2015. The last time was changed the system more (Auzinš et al. 2020a, b)

Source Own elaboration

Estonia has one level of local government. It was short time at the beginning of 1990s when there were two levels of local governments. It was abolished very soon. Further, the county government had the state level and no self-government status. After administrative reform, when the number of local governments decreased, the county government was withdrawn (in 2018). Nevertheless, the spatial plan on the county level remained.

The developer can be a municipality; nevertheless, mostly, it is a private person. The municipal council and officers do not make detailed plans; the developer outsources almost all the preparation work. Still, the local government plays a key role in local planning matters. Among other things, they enact local spatial plans. It should be emphasised that although there are no typical zoning plans in the Estonian spatial planning system, the enacted spatial plans specify the purpose of the land use. In addition, the division between more general plans and specific plans, noticeable in other countries, is replicated. General plans, or in this case comprehensive plans, contain less legally concrete guidelines. They can be likened more broadly to integrated development planning acts. A detailed plan for building activity is required for urban areas such as settlement units, towns, and villages. The adoption of detailed plans in rural areas depends on the judgement of the local authority. If a detailed plan is not mandatory, the comprehensive plan is legally binding. Detailed plans contain more specific guidelines. Also, in Estonia, at the local level, binding graphic parts are part of the plans. Again, one essential element of plans is the demarcation lines between different areas.

Development policy objectives are included in various documents at local level in Estonia. One of the acts containing the indicated elements is the spatial plan (comprehensive spatial plan). In the case under analysis, there is consistency between the different acts (also guaranteed by institutional arrangements). Above the municipalities, there is the county level. At this level, spatial development priorities are defined. These guidelines must be taken into account when developing spatial plans at the local level. There is also a strong relationship in this respect. It should be noted that it is possible to amend the findings at a higher level on the basis of an initial amendment to the findings of spatial plans at the local level. At the national level, there is a national spatial plan synthesising spatial development priorities. It should be emphasised that the planning acts have extensive environmental and natural provisions.

National designated spatial plan prejudges the possibility of locating key (from a national perspective) investments. This solution also contributes to the integration of development policies and different related perspectives. Undoubtedly, this instrument is much easier for smaller countries to prepare. In the case of the larger countries in the study group, some barriers arise to the inclusion of all solutions in a single act.

A problem noticeable in Estonia's local spatial planning practice is the scarcity of a holistic view when shaping detailed plans. Consequently, detailed plans usually follow the perspective of the individual property owner. There are also issues with the implementation of spatial plans. As in the previous case, spatial planning regulations are subject to frequent changes in Estonia as well.

2.6 Spatial Planning in Hungary

Hungary has one of the oldest post-communist planning acts in force across CEE countries, which dates from 1996. Like other spatial planning acts in CEE countries, it has been amended as well, and planning provisions have crossed over into other pieces of legislation. Against this background, the aim of the 2018 act indicated in the table is to harmonise various sectoral aspects (Table 2.4).

There are two levels of local government in Hungary. At the municipal level, there is a legislative authority and an executive (single member) authority. Planning documents are enacted at this level, as well as acts related to development policy. Nevertheless, the role of higher-level local government authorities is also important in this respect. Spatial plans at the local level allow the making of investments. Hence, without a plan, there is no possibility to carry out an investment. This is a major difference from, for example, the Polish spatial planning system. Local spatial plans comprise in-depth development guidelines. They are binding for investors. Drawings are also binding parts of local spatial plan, their main purpose being the demarcation of differentiated areas. Note, also, that the local development plan is part of the local spatial plan. Thus, there is an attempt at integrating integrate spatial planning and development policy fields.

At the regional level, there is the county spatial plan, which is not a binding act for investors. Nonetheless, it does contain detailed spatial planning guidelines that need to be implemented in local spatial planning. This relationship is maintained in practice, as is the relationship between development policy guidelines defined at different levels. At the national level, there is an act that defines spatial development priorities.

A law facilitating and accelerating the implementation of selected investments can be identified as a specific Hungarian solution. This is an example of the attempt to circumvent spatial planning legislation and some of its restrictions. Environmental- and nature-related issues have been developed in an integrated way across different planning levels, with the national-level setting the agenda for translating these issues into secondary legislation.

One Hungarian problem is the way planning regulations are formulated, the challenge here is the adaptation of regulatory specifics to routine planning practices. This lack of adaptation opens the door for investors wishing to challenge these regulations in court. Spatial planning regulations are subject to successive changes and updates.

Table 2.4 A synthetic view of the spatial planning system in Hungary

Information on local government units in the country (which ones are there?). How many units are there at the lowest level (usually municipalities)? What is the area of the country and the population of the country?	In Hungary, there are communes and counties. There are 3155 communes. Hungary has 9689 thousand people, and the surface area is 93,030 km^2 Act Nr CLXXXIX. of 2011 on local governments in Hungary
Brief description of the local level of public authorities. Brief indication of which authorities participate in spatial planning by drawing up instruments (the local level can be diverse. It can be municipalities, but also counties and cities)	Municipalities have a municipal council/body of representatives (*képviselő testület*) (adopts resolutions and decrees—the controlling body) led by the mayor and a mayor's office (*polgármesteri hivatal*)—the executive body. The municipal council adopts local spatial policy tools (*Településterv—Local spatial plan, Építési szabályzat—Building code*) The municipal council/body of representatives issues decisions on development conditions based on the spatial tools adopted by the municipality. Counties elaborate the county regulation plans and development concepts guided by the national spatial tools
Characteristics of legal acts concerning spatial planning	In Hungary, the act Nr XXI of 1996 on regional development and land use planning is in force. Other acts are also important but especially the Act CXXXIX of 2018 on land use plan of Hungary and priority areas which try to harmonise the sectoral issues in spatial planning
Determination of which acts at the local level define the zones (and whether these acts are binding and there is an obligation to enact them)	In Hungary, zoning—land use is determined by local spatial plans (land use means determination of the purpose for which a given area is to be used. It may be, e.g., for housing, services, etc.). Spatial plans are binding acts for investors/property owners. The municipal authorities are obliged to adopt them; without adopted spatial plan, it is not possible to issue building permission
Identify which acts at the local level set out detailed guidelines for land use—development parameters (and whether these acts are binding and there is an obligation to enact them). Identify what these guidelines are	Guidelines for land use are also laid down in the local spatial plans. Spatial plans are binding acts for investors/property owners. The municipal authorities are obliged to adopt them. The local spatial plans set out the following development parameters: building height, building intensity, biologically active area, building line, number of parking spaces, etc. The frames of the rules of local land use regulations are set in the governmental decree Nr. 253/1997. (XII. 20.) on local land use and building regulations The Act LXXIV of 2016 on townscape protection was introduced based on which the municipality issues a local townscape decree and a guideline (amended in 2021. by act Nr. XXXIX. of 2021)
Determine whether a graphic/map is part of the spatial plan at the local level. If yes—is it a binding part	Spatial plans at the local level include a text and a graphic part. The graphic part specifies the intended use of the land and contains the demarcation lines between areas with different uses. The graphic part is also the binding part of the plan
Identify which acts at the local level shape spatial development policy (if different acts, what is more or less in each document)	Development policy at the local level is defined by: • local development plan as part of the local spatial plan

(continued)

Table 2.4 (continued)

	In addition to defining the vision of the municipality, the development plan sets out the short-, medium-, and long-term objectives, tasks, actions, improvements, and measures to be implemented by the municipality in order to ensure its planned, sustainable, and economic development and operation. It contains a detailed development plan which delineates and defines the development action areas as for development and non-development areas in the context of the existing settlement structure of the municipality, with a phasing appropriate to the development objective (Gov. decree 419/2021 (VII. 15.))
Regional-level characteristics	The regional level in Hungary consists of counties (*megye*). There are the following authorities: the county council/board of representatives (which adopts the resolutions) and the board of the county as an executive body
Identify which acts at regional level set out zoning and land use guidelines (or related elements)	The county spatial plan/land use plan defines landscape and environmental protection zones of regional importance, land use system land use guidelines the basic elements of the settlement network, as well as translocal public purpose investments, infrastructure network. It is not a binding act for investors it is realised through the local spatial plans. The national land use plan has a strong influence on county land use plans, it defines the framework of the rules applicable by them, and recent modification gives wider flexibility for the counties
Identify which acts at the regional level define development policy in the spatial dimension	The county development concept is a planning document that provides a basis for the overall long-term development of a country or a region, which defines the long-term, overall development goals of the region, and provides the necessary guidelines and information for the preparation of development programmes for sectoral and related spatial planning and development actors. Basic content is regulated by gov. decree Nr. 218/2009 (X. 6.)
Brief overview of the relationship between acts at regional level and acts at local level	The county development concept must be taken into account when drawing up county spatial development programme and county land use plan and development concept at the local level Spatial development plans of the county must be taken into account when drawing up spatial development plan as part of the local spatial plan at the local level
Identify (if any) which acts at national level relate to spatial planning matters	The Act CXXXIX of 2018 on land use plan of Hungary and priority areas defines the most important spatial issues and frames of spatial plans of lower levels (regional and local)
Identification of specific solutions	'Act on the acceleration and simplification of the implementation of priority investments significant for the national economy', which entered into force in April 2006. The first objective of the act was to 'implement projects financed by the European Union through a faster, simpler and more uniform procedure' but later more and more projects gained priority status. Priority investments are basically subject to procedural benefits under the law. The licencing authority has to examine such applications ahead of non-priority investments and the administrative deadline has been shortened

(continued)

Table 2.4 (continued)

Identify how environmental and nature conservation sections are included in spatial planning acts at local level	The principles of environmental and nature protection are defined: • in law on environmental protection (Environmental Protection Law). But several other laws influence spatial planning such water, forest and soil protection acts. The national spatial plan has also the objective to highlight the interests of these environmentally sensitive issues in spatial planning so specific zones provide protection: zone of forests, zones of ecological network (corridor, core area, buffer zone), water protection zone, zone of arable land with excellent soil conditions, zone of world heritage sites and sites for nomination of world heritage sites. These zones have the objective to provide protection also on local level, but usually just general rules are formed for them on national level The rules of ecological core areas, corridors and the zone of arable land with excellent soil conditions try to limit the possibilities for the designation of development areas, it is just possible with special permission of the chief architect (so actually it is possible). The rules of the zone of world heritage sites oblige the municipalities to integrate the guidelines of the adopted management plan of world heritage site (in case the area of the municipality belongs to one world heritage site) into the elaboration of the framework of local land use system. 95% of the zone of forest from the national plan has to be defined as forest area in the local plan
Problems with spatial plans at local level	Local plans follow the regulations of regional and national plans, but these are sometimes very general and easy to avoid or fulfil without real protection. New regulations provide loopholes such as the regulations related to priority investments
Defining the role of the courts in relation to planning acts at local level	Changes in the spatial regulations (reducing building intensity) can cause financial losses for investors which can turn to court for compensation
Frequency of changes in the spatial planning system	The Nr. XXI. Act of 1996 on Spatial development and land use planning remains in force in its basic scope; however, there were major changes, especially the reform of 2011 which changed the institutional framework of spatial planning when regions (NUTS2) ceased to exist as planning regions. Around EU accession, there was a process which gave a strengthening role for regions in spatial planning but this was reversed completely by the reform of 2011. The reform of 2011 influenced also micro-regional level, 'multipurpose micro-regions' (in the frames which municipalities cooperated in different fields such as social issues and education) ceased to exist now there are districts but these are just administrative regions For land use plans, the act on the national land use plan is highly important, the first one was adapted in 2003 (2008, 2014, 2018), and as it is also a legal regulation, it was amended every 5 years (current law: CXXXIX. of 2018 on land use plan of Hungary and priority areas)

Source Own elaboration

2.7 Spatial Planning in Latvia

In Latvia, there are different laws and regulations dedicated to spatial development planning. It is possible to distinguish distinct laws for the planning system, for the contents of specific documents, as well as separate (general) regulations for planning, land use and building, and the (digital) planning information system. Apart from this dedicated normative framework, there are many additional laws and regulations covering spatial planning issues (Table 2.5).

There are two levels of government (national and local), but there are three levels for spatial development planning in Latvia (national, regional, and local). From a spatial planning perspective, the municipal level is crucial. The position of the legislative authority at this level—which appoints the executive authority—is strong. However, special emphasis should be placed on the role of planning regions (derived public entity having function for planning [I did not get this sentence]), which promote cooperation between planning specialists and influence local spatial planning.

Here, there are three types of spatial plans at the local level: one general and the other two specific. An element of general plans (i.e., local government spatial plans) is 'use of the territory'. Functional zoning is included in the graphic part of the plan. All municipalities must adopt the local government spatial plan. These plans have direct effects on investors. In-depth by-laws for land use and building are contained in local government spatial plans (to a greater extent than, e.g., in general plans in Bulgaria or the Czech Republic). They have a special, separate by-law section. It is the national regulations or general plans that determine the obligation to adopt a detailed plan in an area. Note that this plan is also binding for investors, with drawings being also legally binding. Their key element is the delimitation of individual categories of land. The reason for adopting a detailed plan is primarily due to the serious spatial and construction challenges in that area.

In Latvia, there are strategies at the local level that defines long-term spatial policy objectives. Particularly noteworthy is the inclusion in these acts of functional areas, the definition of the settlement structure, as well as infrastructural and environmental issues. Similar proportional coverage is found in regional-level documents. As far as development policy acts are concerned, there is an optimal correspondence between these different levels. At the national level, there are two main acts for spatial policy objectives, one for land and one maritime, and two other acts including spatial development policies, one comparatively short-term, guiding the implementation of the main spatial perspective for land, and the other one for the special (coastal) area.

Exceptions introduced at the national level should be noted as a specific Latvian solution. These include deemed priority investments that have been excluded from the spatial planning framework. They include renewable energy sources. It is possible to implement some of these investments independently of spatial plans. On the other hand, environmental and nature-related issues, irrespective of the significant scope contained in other laws, are also present in spatial development planning acts.

Numerous problems with spatial plans appear in Latvia. Among other things, they relate to legislation that changes too quickly and or to outdated methods of public participation that cannot meet the rising awareness of spatial planning issues and its need for an in-depth dialogue. One of the main future challenges will be the drafting and enactment of new local government spatial plans for larger municipalities, after the administrative-territorial reform of 2025. Courts play a role in the system mainly when it comes to procedural issues. Nevertheless, their assessments affect spatial planning practices.

Table 2.5 A synthetic view of the spatial planning system in Latvia

Information on local government units in the country (which ones are there?). How many units are there at the lowest level (usually municipalities)? What is the area of the country and the population of the country?	There are 43 local-level municipalities in the Republic of Latvia after administrative-territorial reform 2021: 1. local governments of State cities, in total 7 2. municipality (*novads*) governments, in total 36 (three state cities are included into municipality governments) Latvia has a population of 1.875 million. The surface area of Latvia is 64,589 km^2
Brief description of the local level of public authorities. Brief indication of which authorities participate in spatial planning by drawing up instruments (the local level can be diverse. It can be municipalities, but also counties and cities)	Local governments of state cities and municipalities have elected Councils which approve local governments by—laws, budget, etc.—the decision-making body The Council chairperson manages the work of the Council. Council appoints an executive director—responsible for the work of the local government's institutions, its capital companies Council approves local spatial development planning acts
Characteristics of legal acts concerning spatial planning	In Latvia, there are two main laws which relates to spatial planning—one umbrella law determining planning system covering policy and spatial development planning and second one more directly concerning spatial planning issues such as public participation, competence of public authorities in planning, planning documents and their content at every level—national, regional, and local: • Development Planning System Law, in force from 2009 • Spatial Development Planning Law, in force from 2011 There are several spatial development planning regulations of Cabinet of Ministers—Procedures for the Public Participation in the Development Planning Process, in force from 2009; General Regulations for the Planning, Use and Building of the Territory, in force from 2013; Regulations of the territory development planning information system, in force from 2014; Regulations on local government spatial development planning documents, in force from 2015, etc. Regulations concerning spatial development planning appears also in dozens of other acts such as Regional Development Law, in force from 2002 and law On Environmental Impact Assessment, in force from 1998, Law on Administrative Territories and Populated Areas; Protection Zone Law; Road Traffic Law; Water Management Law; etc., and regulations for carrying out a strategic environmental impact assessment, regulations for nature protection, protection of cultural monuments, etc.
Determination of which acts at the local level define the zones (and whether these acts are binding and there is an obligation to enact them)	In Latvia, land use zones are determined by local government spatial plan (*teritorijas plānojums*), this plan is obligatory, developed for an administrative territory (or its part) In part of a territory local government spatial plan may be amended by local plan (*lokālplānojums*) insofar as the local plan is not in contradiction with the sustainable development strategy of the local government Council approves local government's spatial plan and local plan amending the spatial plan by binding regulations (binding are *Territory use and building regulations* and graphic part of the plan) These plans are legally binding for investors/property owners

(continued)

Table 2.5 (continued)

Identify which acts at the local level set out detailed guidelines for land use—development parameters (and whether these acts are binding and there is an obligation to enact them). Identify what these guidelines are	A local government shall have the following coherently harmonised spatial planning documents: sustainable development strategy, development programme, spatial plan of the local government, local plans, detailed plans, and thematic plans Binding regulations for land use are laid down in the local governments spatial plan, part *Territory use and building regulations* containing requirements for the whole planned territory, main and additional requirements for the use of the territory and building in each functional zone or subzone, and an offer for the development of detailed plans and local plans, etc. Regulations include development parameters such as the minimum new plot of land area; maximum buildings density (%); buildings intensity (%); building height (floors of the house); the minimum free green territories rate (%); other special restrictions for concrete functional land use zone In case of local plan binding regulations, regarding land use and building are in force in the territory of the local plan In several cases, the local government's spatial plans are detailed by detailed plans (*detālplānojums*) developed in order to lay down the requirements for the use and building parameters of specific land units, as well as to adjust the borders of land units and restrictions. The elaboration of detailed plan is mandatory in the cases specified in the spatial plan or the local plan, as well as in cases specified in the regulations (e.g., for the territory in which the construction of the territory complex is planned—a set of buildings, structures, streets, roads, and engineering networks) Detailed plans are approved with a general administrative act (binding for investors/property owners)
Determine whether a graphic/map is part of the spatial plan at the local level. If yes—is it a binding part	Spatial plans at local level (local government spatial plan, local plan, and detailed plan) include binding graphic part—graphic/maps The binding graphic part of the local government spatial plan is functional zoning; territories with special regulations; the burdened territories under the competence of the municipality; village boundaries
Identify which acts at the local level shape spatial development policy (if different acts, what is more or less in each document)	Spatial development policy at the local level is defined by long-term sustainable development strategy, its spatial development perspective Spatial development perspective sets out the prospective spatial structure of the territory, schematically represents the main functional spaces, the most significant elements of the spatial structure and their long-term changes, settlement structures and proposals for the placement of development centres as well as spatial structure of natural territories; main transport corridors and infrastructure; if necessary, the scenically valuable and culturally–historically significant territories; areas to be developed as a priority, etc. It is not binding act for the investors
Regional-level characteristics	The regional level in Latvia consists of five planning regions A planning region is a derived public entity. The competence of it is to ensure regional development planning, coordination, cooperation between municipalities and state administrative institutions. Planning regions elaborate spatial development planning documents—sustainable development strategies and programmes—to be taken into account at the local level Regions have the following authorities: general meeting of the chairmen of municipalities of the planning region, planning region Development Council, consisting of elected representatives from the region's municipalities—decision-making authority, approves the regulations, budget, etc., and region's Administration, which ensures the execution of the decisions taken by the Development Council Region's Administration is supervised by the Ministry of Environmental Protection and Regional Development and is led by the head of Administration

(continued)

Table 2.5 (continued)

Identify which acts at regional level set out zoning and land use guidelines (or related elements)	The region's Sustainable Development Strategy in its spatial development perspective defines main functional spaces and main elements of the regions spatial structure, and the desired long-term changes Strategy is including guidelines, recommendations for the development of the spatial perspectives on the local level, such as guidelines for spatial structure of the settlement, development of traffic infrastructure, spatial structure of rural areas, spatial structure of natural areas, etc. Region's Sustainable Development Strategy is not a binding act for the investors. No act at regional level directly specifies zoning
Identify which acts at the regional level define development policy in the spatial dimension	The region's Sustainable Development Strategy defines development policy in the spatial dimension. It is specifying the vision of the long-term development, strategic objectives, and priorities of the planning region and the spatial development perspective in a written and graphic form Strategy has strategic part—long-term development vision, goals, priorities, and economic profile and spatial development perspective with guidelines—recommendations for the development of spatial perspectives of local governments Spatial development perspective describes and graphically represents desired spatial structure of the region, including the most important existing elements of the spatial structure, the desired long-term changes, the main functional spaces; population structure and development centres; main transport corridors, spatial structure of natural territories, etc.
Brief overview of the relationship between acts at regional level and acts at local level	It is stated in regulations that planning regions Sustainable Development Strategy with its Spatial Perspective is to be taken into account when drawing up municipalities' spatial development planning documents—strategies with spatial perspectives, development programmes, and spatial plans as well as the development programme of a planning region is to be taken into account when drawing up local government's development programme In practice, there are rather optimal relationships between regional and local levels—from one side there are competences of a planning region given by the law to provide opinions on the conformity of the draft local government strategies and development programmes with the spatial development planning documents of the region and from the other side the elaboration process of region's planning documents is in collaboration between regional administration and municipal planning specialists, representatives of the state administration and the non-governmental sector, and experts. No less important that regional-level planning documents are adopted by the local government politicians who represent municipalities at the region Development Council
Identify (if any) which acts at national level relate to spatial planning matters	To spatial planning matters at the national level relate long-term acts—Sustainable Development Strategy of Latvia until 2030 (Latvia2030), with its Spatial development perspective covering issues such as improvement of accessibility, settlement, spaces of national interest, and Maritime Spatial Plan of Latvia—spatial development planning document that defines the use of the sea space Two other national-level acts relating to some spatial planning matters are middle-term National Development Plan of Latvia, for 2021–2027 giving directions for the spatial development to spaces of national interests defined in Latvia2030 (NDP2027 Spatial Development Perspective) and National long-term thematic plan for the development of public infrastructure in the coastal area of the Baltic Sea—first national-level thematic plan The policies and solutions of upper levels relating to spatial planning matters translate into spatial development planning at lower levels

(continued)

Table 2.5 (continued)

Identification of specific solutions	In Latvia, there are exception in the entire spatial planning order—objects of national interest—territories and objects necessary to ensure essential public interests, protection, and sustainable use of natural resources (e.g., Rail Baltic line, military objects, etc.) 29 September 2022 the Law on simplified procedures for the construction of wind power plants for the promotion of energy security and independence about simplified procedure for the construction of wind power plants with a total capacity of at least 50 MW as well as the constructions of solar panels (at least 10 MW) was adopted by Latvian Parliament (Saeima). The law stipulates that the construction of wind power plants may qualify for the status of national interest. By the law, wind parks and solar parks are allowed outside towns and villages in the land use zones of industrial construction, technical construction and agricultural areas specified in the local governments spatial plan as well as wind parks are permitted in the forestlands This means that investments in national interest objects can be implemented independently from the local planning acts (e.g., local government spatial plan) and wind and solar parks will be implemented in land use zones stated by the above-mentioned special law
Identify how environmental and nature conservation sections are included in spatial planning acts at local level	The principles of environmental and nature protection are defined in laws, such as Environmental Protection Law, law On Specially Protected Nature Territories and several regulations of Cabinet of Ministers', such as Procedures for Carrying Out a Strategic Environmental Impact Assessment, General Regulations on Protection and Use of Specially Protected Nature Territories Spatial development planning documents on local level shall take into account appropriate content and requirements of those acts and others of this field, environmental and nature conservation issues are included in the planning acts. Deeper evaluation of environmental situation is in the cases when strategic environmental impact assessment (SEIA) Environmental Report is prepared In several cases, there are observed discrepancies, contradictions between the requirements of the nature conservation plans developed for national parks, other valuable Natura2000 areas, and local governments spatial plans (lack of joint view on the development by nature plan developers and locals). Some projects have been carried out to ensure mutual integration of these plans
Problems with spatial plans at local level	Problems with spatial plans at local level are that different laws and regulations, which contain rules for spatial planning, change too often, the methods of public participation used do not provide for in-depth dialogue even though awareness—raising is increasing and the provisions of spatial plans are often not implemented Future challenge will be the elaboration and adoption of new local government spatial plans for larger-scale municipalities after the administrative-territorial reform by 2025 using the same legislation and detailing. Part of local governments, lacking appropriate human resources, are outsourcing the elaboration of the plans to private consultant companies, challenge will be their capacity for this amount of work in short time period, etc.

(continued)

Table 2.5 (continued)

Defining the role of the courts in relation to planning acts at local level	Two courts: Constitutional Court and Administrative Court planning practice at local level in Latvia Local government's spatial plan or local plan (issued as regulatory acts-binding regulations of municipality) may be challenged by property owners to Constitutional Court after the Ministry of Environmental Protection and Regional Development has evaluated their submissions and if violations of the procedures for developing the plan or a non-conformity with the requirements of laws and regulations are established. Constitutional Court evaluates the compliance of the planning document binding regulations with legal norms of higher legal force, including fundamental rights and general principles of law and planning, thus providing an explanation and forming legal norms in practice and the application of principles in planning The detailed plan is appealed to the Administrative Court. The jurisprudence of the Administrative Court influences and shapes the practice of the implementation of detailed plans
Frequency of changes in the spatial planning system	The Development Planning System Law, 2009, and Spatial Development Planning Law, 2011, and regulations for spatial development planning elaborated after remain in force in their basic, unchanged scope Since the adoption of these two main laws regulating the spatial planning system there have been smaller or major detailed changes but without changing the basic principles and instruments of these laws

Source Own elaboration

2.8 Spatial Planning in Lithuania

Lithuanian laws relate primarily to territorial planning. The laws are very diverse in this case—there are as many as 36 different regulations. This diversity causes very serious interpretation problems. This type of arrangement is typical for a large part of the CEE countries, we have studied (Table 2.6).

There are two levels of local government in Lithuania, with the local level being crucial for spatial planning. This is also reflected administratively. At the municipal level, there are two types of authorities: legislative and executive (single member). From an organisational point of view, the preparation of draft plans is coordinated by the executive authority of the municipality. The initiative to proceed with the plans and the subsequent adoption of these plans rests with the legislative authority. There is no classic separation of zones—in this case, there is a certain peculiarity in Lithuanian solutions. Instead, one can point to the delimitation of functional zones, undertaken at different levels by the public authority. Nonetheless, the actual purpose of land use determines the specific division of the plot, in a bottom-up fashion [This sentence is not entirely clear]. Specific plans are linked to urban or urbanising areas. These plans contain both general and specific guidelines (hence, they act as a counterpart of the general and detailed plans in Bulgaria and the Czech Republic). Thus, in the Lithuanian case, general plans are more like strategic documents. Drawings are also an integral part of these plans. Priority development territories are a separate part of spatial plans. It is the spatial plans that broadly define development policies. Other planning documents at local level play a far less prominent role.

At the regional level, there are also documents explicitly referring to spatial planning. However, some spatial planning elements appear in regional comprehensive plans. Particular attention is paid to the coherence of development policy acts at different levels (which translates to a lesser extent into spatial planning). At the national level, there is the Comprehensive Plan of the Territory of the Republic of Lithuania, which is universally binding. It sets out differentiated guidelines, both for acts at the lower levels and for sectoral acts adopted at the central level. The environmental and nature dimensions are considered in spatial planning as well, with the scope of action at the discretion of local authorities. The role of courts is reduced to verifying related administrative decisions in spatial planning. Relatively frequent modifications of spatial planning regulations are present.

2.9 Spatial Planning in Poland

The Spatial Planning and Development Act has been in force in Poland since 2003. The law contains numerous inconsistencies, thereby causing serious dilemmas when interpreting its provisions. Furthermore, spatial planning issues spread across dozens

Table 2.6 A synthetic view of the spatial planning system in Lithuania

Information on local government units in the country (which ones are there?). How many units are there at the lowest level (usually municipalities)? What is the area of the country and the population of the country?	The territorial administrative units of the Republic of Lithuania are counties and municipalities. The territory of the Republic of Lithuania currently comprises 10 counties and 60 municipalities which consist of 546 elderships The county is a higher administrative unit. It is formed from the territories of the municipalities characterised by common social, economic, and ethno-cultural interests Lithuania has a population of 2,830,097 (01 July 2022) The surface area of Lithuania is 65.286 km^2
Brief description of the local level of public authorities. Brief indication of which authorities participate in spatial planning by drawing up instruments (the local level can be diverse. It can be municipalities, but also counties and cities)	Municipalities have a municipal council (a representative institution of the municipality) and the director of the municipal administration (an executive institution of the municipality) The preparation of the comprehensive plans of a municipality or parts thereof shall commence upon adoption by a municipal council of a decision on the commencement of preparation of the comprehensive plans of the municipality or parts thereof and planning objectives The preparation of such plans shall be organised by the director of municipal administration The director of municipal administration shall submit the comprehensive plan of a municipality or the comprehensive plan of a part of the municipality for approval to the municipal council The preparation of documents of municipal-level and local-level territorial planning shall be organised by the director of municipal administration Counties are only territorial administrative units that do not have their own government since 2010

(continued)

Table 2.6 (continued)

Characteristics of legal acts concerning spatial planning	In Lithuania, current legal acts concerning territorial (spatial) planning: • Law on Territorial Planning of the Republic of Lithuania • Law of the Republic of Lithuania on Territorial Planning and the Law on State Supervision of Construction 36 different Regulations Issues concerning spatial planning appear in many other acts concerning other issues (e.g., Law of the Republic of Lithuania on the Protected Areas, Law on the Protection of Immovable Cultural Heritage, etc.) Because of that, many disputes and problems arise in the spatial planning process
Determination of which acts at the local level define the zones (and whether these acts are binding and there is an obligation to enact them)	In Lithuania, functional zoning shall mean the division of a territory into zones establishing the priorities of the principal designations of the use of land and possible activities therein, which is carried out while preparing territorial planning documents based on the levels and tasks of the planning in compliance with the Rules for the Preparation of Documents of Complex Territorial Planning approved by the Ministry of Environment of the Republic of Lithuania (hereinafter: the 'Ministry of Environment') Zoning-land use can be determined by projects of formation and rearrangement of land parcels (they are not spatial planning documents)
Identify which acts at the local level set out detailed guidelines for land use—development parameters (and whether these acts are binding and there is an obligation to enact them). Identify what these guidelines are	Detailed plan shall mean a document of local-level complex territorial planning of an urbanised territory or a territory under urbanisation which establishes regulations on the use of the territory Detailed plans shall be prepared in urbanised territories and territories under urbanisation as established in municipal-level or local-level comprehensive plans, if prepared, where development of the territory is envisaged and/or where the regulations on the use of the territory are amended in urbanised territories and territories under urbanisation When detailing a municipal-level comprehensive plan and/or a local-level comprehensive plan, if prepared, in compliance with the Rules for the Preparation of Documents of Complex Territorial Planning, the following regulations on the use of a territory, mandatory for natural and legal persons or other organisations, shall be established in detailed plans: 1. The type of use of the territory conforming to the local-level comprehensive plan, if prepared, or a specific principal designation of land use and specific modes of land use 2. The permitted height of buildings 3. The permitted density of development of land parcels 4. The permitted intensity of development of land parcels or density of development volume (in territories of industrial and warehousing facilities and/or engineering infrastructure) 5. Possible types of development conforming to the local-level comprehensive plan, if prepared, the construction zone, boundary, and line of construction works 6. Boundaries of territories necessary for the social and engineering infrastructure and/or engineering communications corridors 7. Possible boundaries of land parcels and/or principles of forming and reparcelling of land parcels (minimum and/or maximum possible sizes of land parcels) 8. Separate green areas, parts of territories of subordinate green areas, and of greenery in percentage

(continued)

Table 2.6 (continued)

	Taking into account the scale of the detailed plan under preparation, the planned territory, its value and planning tasks formulated in the programme of planning works, the organiser of planning shall indicate in the programme of planning works that additional legislation-based mandatory requirements relating to protection of the environment, landscape, natural and immovable cultural heritage (sites of cultural heritage and their protection zones, objects of cultural heritage, their territories and protection zones) and regulations relating to urbanism and architecture (layout of buildings and public spaces, layout of parking places, etc.), development of the engineering and social infrastructure, public health, or other regulations on the use of the territory shall be established in the detailed plan Specific modes of land use in a land parcel or a part thereof based on the type of use of the territory approved in a detailed plan shall be established when approving the detailed plan, and they shall be replaced with other possible modes of land use by a decision of the director of municipal administration, providing information to the public in accordance with the procedure established by the government
Determine whether a graphic/map is part of the spatial plan at the local level. If yes—is it a binding part	Documents of complex (comprehensive and detailed plans) and special territorial planning include graphic and written form, solutions of the use, management and protection of territories, and the needs and conditions for development thereof. The graphic part is also the binding part of the plan
Identify which acts at the local level shape spatial development policy (if different acts, what is more or less in each document)	Priority development territories shall mean the territories (or parts thereof) urbanised or under urbanisation identified in municipal-level and local-level comprehensive plans, in which a municipality undertakes to develop the social and/or engineering infrastructure Various development programmes, research and feasibility studies, surveys, project proposals representing proposals on the development of activities or restrictions on certain activities in a specific territory and justification therefor or detailing or justifying the solutions of territorial planning documents shall not constitute territorial planning documents
Regional-level characteristics	The Law on Regional Development defines the general framework for regional policy, including urban development. Its overarching objective is to reduce disparities across and within regions and to promote balanced and sustainable development throughout the country This policy is implemented by the Regional Development Councils and National Regional Development Council
Identify which acts at regional level set out zoning and land use guidelines (or related elements)	Territorial planning documents are not prepared for regions In compliance with the Rules for the Preparation of Documents of Complex Territorial Planning, the following mandatory provisions for the use of a territory, which are consistent with and detail the comprehensive plan of the territory of the State, shall be established for state and municipal institutions in the comprehensive plans of a part of the territory of the State: Functional zoning of the use of the part of the territory of the state; the system of state parks, strict state reserves and state reserves, the system of state-protected sites and sites of cultural heritage of national significance and regional significance and the system of territories of complex objects of cultural heritage and protection zones thereof; etc.
Identify which acts at the regional level define development policy in the spatial dimension	The tasks of the comprehensive plan of a part of the territory of the state shall be to establish guidelines for the implementation of territorial development policy of the Republic of Lithuania in separate parts of the territory of the State, the spatial structures of these territories and elements thereof The tasks of municipal-level and local-level comprehensive plans shall be to form the directions of functional and spatial development of a territory consistent with the level of planning

(continued)

Table 2.6 (continued)

Brief overview of the relationship between acts at regional level and acts at local level	The organisational framework of application of territorial planning documents and their interaction shall operate in compliance with the law of territorial planning and other legal acts. Each lower level of territorial planning must operate in compliance with the approved solutions of documents of higher-level territorial planning and detail them. The solutions of documents of same level territorial planning must be coordinated
Identify (if any) which acts at national level relate to spatial planning matters	The Comprehensive Plan of the Territory of the Republic of Lithuania is one of the central and integrated territorial planning document's which is executed on the state level; establishes the spatial structure of a planned territory and mandatory provisions and requirements for the use of the territory as well as the principles of its protection The Comprehensive Plan of the Republic of Lithuania is a legally binding document: • obligatory for the state governmental institutions, taking decisions on national level, related to the use, management, and protection of the territory of the country, forming regional policy, spatial integrated policy • forms planning conditions for national-level special plans, long-term programmes and strategies, lower-level comprehensive and special plans • for the development of strategies of economic sectors, other strategic plans and programmes of state institutions have to rest upon solutions of the Comprehensive Plan of the Republic of Lithuania
Identification of specific solutions	The Territorial Planning Law of the Republic of Lithuania provides for the specific legal status of territorial planning documents for projects important to the state. The solutions of documents of territorial planning of projects of importance to the State shall be of higher legal validity than the solutions of documents of municipal-level and local-level complex and special territorial planning and shall apply on a mandatory basis to municipalities when preparing, amending, or adjusting the documents of municipal-level and local-level territorial planning. The solutions of the documents of territorial planning of projects of importance to the State shall be mandatory for documents of state- and lower-level territorial planning. When applying the solutions of the documents of territorial planning of projects of importance to the State, the solutions of the documents of state- and lower-level territorial planning shall apply to the extent they do not contradict the solutions of the documents of territorial planning of projects of importance to the State. Taking into account the content of a project, documents of territorial planning of projects of importance to the State shall be prepared as either documents of complex territorial planning or documents of special territorial planning
Identify how environmental and nature conservation sections are included in spatial planning acts at local level	When establishing planning objectives for a specific territory, it is necessary to take account of the public needs, the landscape and biodiversity of a planned territory, its geographical location, geological conditions, the existing urban, engineering, transport, and agricultural systems, the interests and rights of managers and users of land and other immovable property and third parties, as well as architectural, environmental, public health, nature protection and heritage protection requirements and state and public security and defence needs How environmental and nature protection sections are included in territorial planning acts at the local level is basically at the discretion of the municipalities
Problems with spatial plans at local level	In Lithuania, the problems of territorial planning at the local level are related to the incompatibility of various interests (investors and the local community) or restrictions applied by the owner

(continued)

Table 2.6 (continued)

Defining the role of the courts in relation to planning acts at local level	Administrative decisions unlawfully adopted during the process of territorial planning (before the approval of the territorial planning document) shall be revoked by the entities which have adopted those decisions or the court at the request of the persons concerned or on their own initiative in accordance with the administrative procedure. The right to apply to the institution which has adopted the decisions or the court shall be granted to persons whose rights and legitimate interests have been violated, also, on the basis of complaints or reports of the said persons or on their own initiative, to the institutions carrying out state supervision of territorial planning within their remit, other entities of public administration, representatives of the public concerned in defence of the public interest An unlawfully adopted decision on the approval of a territorial planning document shall be revoked by the court. The court may adopt an administrative decision to revoke the full territorial planning document or a part thereof The right to apply to court for the revocation of an unlawfully adopted administrative decision on the revocation of approval of a territorial planning document shall be granted to the institutions carrying out state supervision of territorial planning and, within their remit, other entities of public administration whose rights and legitimate interests have been violated Representatives of the public concerned, other natural and legal persons concerned shall have the right to apply to court regarding the adopted administrative decision on the revocation of approval of a territorial planning document only in the cases where: 1. they have been involved in the publicity procedures of territorial planning and have lodged complaints or reports regarding the decisions relating to territorial planning adopted by entities of public administration or regarding the failure to act of these entities to the institutions carrying out state supervision of territorial planning before the adoption of the appealed administrative decision on the approval of the territorial planning document, where in the course of the publicity procedures they were aware of or could have objectively foreseen the possible violation of their rights 2. due to reasons recognised by the court as serious, they could not have been involved in the publicity procedures of territorial planning and lodged complaints or reports regarding the decisions relating to territorial planning adopted by the entities of public administration or regarding the failure to act of these entities to the institutions carrying out state supervision of territorial planning before the adoption of the appealed administrative decision on the approval of the territorial planning document
Frequency of changes in the spatial planning system	After Lithuania had regained its independence, the issues related to territorial planning were in part regulated in a single legislative act—the Interim Regulations on Territorial Planning of the Republic of Lithuania, approved by the Decree of the Government of the Republic of Lithuania of 1993-03-12 The legislation of territorial planning in Lithuania had developed very intensively since the first stage of the modernisation, when the first Law on Territorial Planning of the Republic of Lithuania was adopted on 1 January 1996 The third and fundamental stage of the territorial planning reform was completed in 2014. But, even after the third reform, this legislative enactment had been amended and supplemented 21 times

Source Own elaboration

of other acts. Obviously, significant discrepancies between these laws are a serious problem (Table 2.7).

There are three levels of local administration in Poland. At the local level, municipalities have both a legislative authority and an executive authority. The executive initiates and then enacts spatial planning acts. The legislature, on the other hand, is responsible for preparing specific projects. Interference by higher-level local authorities in local spatial planning takes place but is very limited.

In Poland, there is one type of spatial plan at the local level. It consists of both text and drawings, which together specify, as the Polish legislator puts it, 'land use'. Spatial plans are binding for investors when enacted, but municipal authorities do not have to adopt them. Against this background, the greater part of Poland's territory is not covered by such plans. This is a very peculiar situation even compared to other CEE countries. Local spatial plans set out both general and detailed guidelines for development and land use. Nonetheless, they are fewer in number than in other CEE countries. There are also drawings in these plans that show the demarcation between areas of a different use. They are also binding for investors.

At the local level, there are development strategies and spatial planning studies. The distinction between the two is not entirely clear. Development strategies are typically strategic documents. Nevertheless, they contain certain guidelines for local spatial planning. Spatial planning studies incorporate more detailed guidelines that need to be considered within spatial plans. This solution only works to a limited extent, while at the same time leading to numerous conflicts and barriers to implementation.

A similar pattern exists at the regional level. The spatial development plans of the provinces are the counterparts of the spatial planning studies. There are also regional development strategies. According to the literal wording of the laws currently in force, all these documents should be interdependent and compatible. In practice, however, the effects they produce are very limited in scope. Consequently, the regional level determines the direction of local spatial and development policy rather poorly. At the national level, there is currently no act that addresses spatial planning and development issues comprehensively.

Table 2.7 A synthetic view of the spatial planning system in Poland

Information on local government units in the country (which ones are there?). How many units are there at the lowest level (usually municipalities)? What is the area of the country and the population of the country?	In Poland, there are communes, districts, and voivodships. There are 2489 communes. Poland has a population of 37.95 million. The surface area of Poland is 322,575 km^2
Brief description of the local level of public authorities. Brief indication of which authorities participate in spatial planning by drawing up instruments (the local level can be diverse. It can be municipalities, but also counties and cities)	Municipalities have a municipal council (adopts resolutions—the controlling body) and a mayor/mayor (depending on the municipality)—the executive body. The municipal council adopts local spatial policy acts (studium uwarunkowań i kierunków zagospodarowania przestrzennego, miejscowe plany zagospodarowania przestrzennego). The mayor issues decisions on development conditions. Counties are not more extensively involved in spatial planning

(continued)

Table 2.7 (continued)

Characteristics of legal acts concerning spatial planning	In Poland, the Act of 27 March 2003 on spatial planning and development is in force. However, issues concerning spatial planning appear in dozens of other acts concerning other issues (e.g., environmental protection, nature protection, and protection of monuments). This often causes interpretation problems
Determination of which acts at the local level define the zones (and whether these acts are binding and there is an obligation to enact them)	In Poland, zoning land use is determined by local spatial plans (land use means determination of the purpose for which a given area is to be used. It may be, e.g., for housing, services, etc.). Spatial plans are binding acts for investors/property owners. The municipal authorities are not obliged to adopt them. It depends on their discretion
Identify which acts at the local level set out detailed guidelines for land use—development parameters (and whether these acts are binding and there is an obligation to enact them). Identify what these guidelines are	Guidelines for land use are also laid down in the local spatial plans. Spatial plans are binding acts for investors/property owners. The municipal authorities are not obliged to adopt them. It depends on their discretion. The local spatial plans set out the following development parameters: building height, building intensity, biologically active area, building line, number of parking spaces
Determine whether a graphic/map is part of the spatial plan at the local level. If yes—is it a binding part	Spatial plans at the local level include a text and a graphic part. The graphic part specifies the intended use of the land and contains the demarcation lines between areas with different uses. The graphic part is also the binding part of the plan
Identify which acts at the local level shape spatial development policy (if different acts, what is more or less in each document)	Development policy at the local level is defined by: • local development strategies • studies of spatial development conditions and directions (spatial development plan) The municipality's development strategy sets out the strategic spatial objectives, the expected results of activities, a model of the functional-spatial structure, as well as arrangements and recommendations for shaping and implementing the spatial policy (binding for the studies of spatial development conditions and directions) The study of spatial development conditions and directions defines the directions of changes to the spatial structure (binding for local spatial plans). It is not a binding act for investors
Regional-level characteristics	The regional level in Poland consists of voivodeships. There are the following authorities: the voivodship parliament (which adopts the resolutions of the voivodship) and the board of the voivodship as an executive body (headed by the marshal of the voivodship)
Identify which acts at regional level set out zoning and land use guidelines (or related elements)	The voivodeship spatial development plan defines the basic elements of the settlement network of the voivodeship and their connections as well as translocal public purpose investments. It is not a binding act for investors No act specifies zoning or land use guidelines
Identify which acts at the regional level define development policy in the spatial dimension	The voivodeship development strategy specifies the strategic objectives in the spatial dimension, the expected results of actions and indicators for their achievement, a model of functional and spatial structure on a voivodeship scale and arrangements and recommendations for the implementation of spatial policy on a voivodeship scale

(continued)

Table 2.7 (continued)

Brief overview of the relationship between acts at regional level and acts at local level	The voivodeship development strategy must be taken into account when drawing up voivodeship spatial development plans and development strategies at the local level Spatial development plans of the voivodeship must be taken into account when drawing up studies of spatial development conditions and directions at the local level It is a very weak relationship
Identify (if any) which acts at national level relate to spatial planning matters	There is currently no such act in Poland To a limited extent, issues related to spatial planning are included in national development documents This does not translate into spatial planning at lower levels
Identification of specific solutions	In Poland, the so-called specustawy can be considered a specific solution. These are acts concerning specific types of investments (e.g., road investments, railway investments), independent from the entire spatial order. This means that investments specified in specustawy can be implemented independently from basic planning acts (e.g., local spatial plans)
Identify how environmental and nature conservation sections are included in spatial planning acts at local level	The principles of environmental and nature protection are defined: • in laws on environmental protection (Environmental Protection Law) and nature conservation (Nature Conservation Law) • resolutions concerning forms of nature protection (national parks, nature reserves, etc.) Studies of spatial development conditions and directions and local spatial plans shall take into account all the contents of those acts. Failure to take them into account creates the possibility of annulment of the acts Additional elements created by the local spatial plans themselves will be restrictions on development in environmentally and naturally valuable areas. The scope of restrictions largely depends on the discretion of the municipalities adopting the plans. This applies especially to areas which are not forms of nature conservation (with respect to the latter most restrictions are included in their acts)
Problems with spatial plans at local level	In Poland, a key issue is the economic consequences of a spatial plan. If a spatial plan introduces restrictions in a particular area and leads to a reduction in the value of a property, the owner of such property may demand financial compensation from the municipality. This creates a risk of serious financial losses for municipalities and discourages them from adopting spatial plans When a spatial plan is not in force in a given area, its equivalent becomes an administrative decision: the decision on land development conditions. It is issued by the mayor at the request of the investor. The criteria for issuing this decision are very subjective
Defining the role of the courts in relation to planning acts at local level	Studies of spatial development conditions and directions and spatial plans (much more frequently) may be challenged by property owners before administrative courts. The courts assess not only procedural issues, but also the content of the acts (e.g., they assess whether the scope of land use restrictions is justified). This involves administrative courts to a very large extent in shaping spatial policy
Frequency of changes in the spatial planning system	The 2003 Law on Spatial Planning remains in force in its basic, unchanged scope. However, both this Act and other Acts (containing spatial planning regulations) have been modified in detail. Since 2003, there have been more than a dozen such major amendments (but without changing the basic principles and instruments)

Source Own elaboration

A special case of Polish solutions consists of the so-called specustawy. They are acts that deal with investments that can be implemented against the findings of spatial plans. In such cases, the legislator explicitly indicates the possibility of ignoring planning regulations, resulting in a very broad possibility of ignoring the spatial planning perspective.

Environmental and natural issues are included in Polish planning acts. Other acts and guidelines must also be considered in local spatial plans. Nevertheless, municipal authorities have the possibility to introduce certain restrictions on development. The justification for these restrictions is generally based on environmental and natural reasons.

There are many problems within the Polish spatial planning system: spatial chaos, uncontrolled suburbanisation, an inadequate recognition of the public interest and its translation into planning practice, as well as legal and procedural barriers [against?]. All these issues require regular amendments to spatial planning regulations, as well as numerous attempts to introduce new laws on spatial planning.

Courts also play a very important role in the Polish system. They examine both complaints against planning acts (spatial plans, but also studies on spatial development conditions and directions) and against administrative decisions on spatial planning. In practice, this often translates into serious interference by the courts in local planning decisions.

2.10 Spatial Planning in Romania

The current Romanian planning system has developed in stages. In 1989, the Socialist Planning Act of 1974 was repealed, and with it the entire normative framework. A legal vacuum ensued, with some basic planning provisions absorbed in the Building Act of 1991. The current Spatial Planning Act dates from 2001, with its statutory instruments (methodological norms) enacted only in 2016, but elements related to spatial planning spread across other laws as well (Table 2.8). Both planning and building legislation have undergone repeated revisions. As a result, legal texts are sometimes difficult to follow.

The main planning instruments in Romania are regional and urban plans, which are, essentially, spatial plans. Provisions within regional plans are only binding for the public administration (i.e., steering), whereas urban plans are universally binding (i.e., regulatory).

In Romania, there are two levels of local administration. At the upper level, there are counties, and at the lower, municipalities. Municipalities vary greatly in size and in complexity, a case in point being Bucharest, which has a special status among local administrative units (LAU2). This makes comparisons somewhat difficult, while at the same time providing for an interesting case study that requires a separate analysis. These differences notwithstanding, it is at the local level that the key spatial planning acts are adopted. At a higher level, counties are seriously involved in shaping supra-local (regional) spatial and development policy.

Zoning is defined in General Urban Plans, which are legally binding. Both the General Urban Plans and the Zonal Urban Plans contain rather comprehensive guidelines that define the possibilities for development in a specific area. In Romania, the urban plans also include drawings that set out the different zoning areas.

At the local level, it is possible to distinguish between different acts addressing development policy issues. To a certain extent, they must provide a reference point for local spatial policy. At the regional level, spatial policy elements appear more prominently, especially in County Spatial Development Plans. Note, however, that provisions within these plans are only mandatory for the public administration. Furthermore, the technical hierarchy of plans at different does not fully translate into a symmetrical legal hierarchy. This is a fact often overlooked by planning practitioners.

In the case of Romania, environmental and natural issues are present in spatial planning acts. However, they are usually imported from provisions originating in other laws or norms.

Problems within the Romanian spatial planning system include the poor translation of spatial planning regulations into other fields. Furthermore, provisions are sometimes far too vague for an operative urban planning practice. Planning restrictions can be challenged in court. Consequently, the courts also exert influence on the direction of local spatial policies. Spatial planning regulations are subject to frequent change.

Table 2.8 A synthetic view of the spatial planning system in Romania

Information on local government units in the country (which ones are there?). How many units are there at the lowest level (usually municipalities)? What is the area of the country and the population of the country?	Romania is divided into communes (*comune*), towns (*orașe*), cities (*municipii*), and counties (*județe*) At present, there are 2685 communes, 181 towns, 82 cities, and 41 counties. The capital, Bucharest, has similar status to a county Romania has a resident population of some 19 million. The surface area of Romania is 238,391 km^2
Brief description of the local level of public authorities. Brief indication of which authorities participate in spatial planning by drawing up instruments (the local level can be diverse. It can be municipalities, but also counties and cities)	Communes, towns, and cities have a local council (*consiliu local*), which acts as a deliberative body, and a mayor (*primar*) that leads the executive body The local council adopts urban planning and urban strategy documents. The mayor issues decisions on development conditions Similarly, counties have an elected deliberative body, the county council (*consiliul județean*) that adopts various resolutions, and an executive body headed by an elected county council president (*președintele consiliului județean*) The county council adopts regional and urban planning documents, as well as spatial strategy documents where more than one local territorial administrative unit (LAU2) within the county boundaries is involved

(continued)

Table 2.8 (continued)

Characteristics of legal acts concerning spatial planning	Law no. 50 of 1991 (*Legea 50/1991, privind autorizarea executării lucrărilor de construcții*), the Building Permission Act, deals with the process of authorising construction works, conditioning it upon the existence of a building permit. The building permit requires, among others, a valid planning certificate, which is issued from an approved urban plan or, in rare circumstances, from an approved spatial (i.e., regional) plan Law no. 350 of 2001, the Urban and Regional Planning Act (*Legea 350/2001, privind amenajarea teritoriului și urbanismul*) regulates urban and spatial activities and specific procedures, specifying the roles of institutional actors, including specific plans and documents, in drafting, approving, and endorsing urban and spatial plans Furthermore, the Government Decision no. 525 of 1996, essentially the General Zoning Regulations (*Hotărârea 525/1996 pentru aprobarea Regulamentului general de urbanism*), states the regulations applicable to all urban plans In addition to these three laws, there are other important laws dealing with different planning matters
Determination of which acts at the local level define the zones (and whether these acts are binding and there is an obligation to enact them)	In Romania, zoning is determined by General Urban Plans (*planurile urbanistice generale/PUG*). They include the delimitation of land use areas, as well as the rules for each zone Restrictions may include authorised and unauthorised functions, easements, as well as general morphological indicators relating to plot characteristics, to their building capacity, and to the buildings' positions on a given lot General Urban Plans are legally binding acts
Identify which acts at the local level set out detailed guidelines for land use—development parameters (and whether these acts are binding and there is an obligation to enact them). Identify what these guidelines are	A more recent practice allows Master Plans to establish guidelines for land use. Guidelines are non-binding, until binding spatial planning documents adopt them as rules However, the local executive body may use these guidelines to recommend the rejection of investment proposals The deliberative body usually adopts the resolutions based on the executive body's recommendations, even though they are not binding At present, the Romanian planning system uses two types of plans: regional (spatial) (*planuri de amenajarea teritoriului*) plans and urban plans (*planuri urbanistice*) The main difference between them is that spatial plans are steering in character (or internally binding, in the sense that they are only mandatory for the public administration), whereas urban plans are regulatory in character, in the sense that they are legally binding for everyone There is, however, the peculiar case of the General Urban Plan (*planul urbanistic general/PUG*), which is both steering and regulatory in character In addition, provisions contained within higher-tier regional plans are mandatory for both lower-tier regional plans and urban plans. However, their provisions are only mandatory for the public administration There are two types of land use and zoning plans in the Romanian planning systems: the General Urban Plan (*planul urbanistic general/PUG*) and the Zonal Urban Plan (*planul urbanistic zonal/PUZ*). Both are urban plans and hence regulatory in character (i.e., their provisions are mandatory for everybody, not only for the public administration) Provisions within a General Urban Plan (*planul urbanistic general/PUG*) are classified into two categories: short-term provisions on the one hand, and mid- and long-term provisions, on the other. Land use and zoning regulations pertain to the first category of provisions, i.e., short-term provisions. The short-term provisions within the General Urban Plan are related to:

(continued)

Table 2.8 (continued)

	1. Contours of the built-up area, in relation to the administrative area of the municipality 2. Land use regulations for the built-up area 3. The zoning pattern, in relation to the transportation network 4. Areas affected by public easements 5. Modernisation [overhaul] and development of technical and municipal infrastructures 6. Protected areas and protection areas for historical monuments and archaeological sites 7. Special protection areas 8. Forms of ownership and legal circulation of land 9. Siting, arrangement, and form conditions for built and planted volumes [bulk] 10. Natural risk areas and their associated risk prevention measures for land use and building 11. Man-made risk areas due to historical landfills The Zonal Urban Plan (*planul urbanistic zonal/PUZ*) follows a somewhat similar structure of provisions, albeit a less ambitious one: 1. Organisation of the road network 2. The architectural and urban layout of the intervention area, in accordance with the urban [city] structure 3. Land use 4. Municipal infrastructure [public utilities] 5. Legal status of plots and land transfers [i.e., market transactions, exchanges, and eminent domain] 6. Historical monument [built heritage] protection and public easements that affect their protection areas
Determine whether a graphic/map is part of the spatial plan at the local level. If yes—is it a binding part	Urban plans include a drawings and text. Drawings specify intended land uses and their demarcation lines Texts include substantiation (background) studies, technical notes, zoning and building regulations, as well as the action plan for implementing the plan
Identify which acts at the local level shape spatial development policy (if different acts, what is more or less in each document)	Development policies at the local level include: • Spatial development strategies • Urban mobility strategies • Economic development strategies • Development guidelines Local development policies set out the strategic objectives, the measures that must be taken to reach these objectives, and the expected results of activities Development policies are non-binding acts but are used as guidelines when developing binding spatial planning documents
Regional-level characteristics	In Romania, the only regional level with an administrative role is the counties (NUTS3 regions) Counties have an elected deliberative body, the county council, that adopts various resolutions, and an executive body, headed by an elected county council president
Identify which acts at regional level set out zoning and land use guidelines (or related elements)	The zoning set out by the County Spatial Development Plan (*planul de amenajare a teritoriului județean/PATJ*), which is essentially a regional plan, has only a strategic purpose However, its provisions are only mandatory for the public administration; i.e., they have a steering character

(continued)

Table 2.8 (continued)

Identify which acts at the regional level define development policy in the spatial dimension	The County Spatial Development Plan defines the spatial development strategy of the county When its provisions are integrated into urban plans (*planuri urbanistice*), they become legally binding
Brief overview of the relationship between acts at regional level and acts at local level	In Romania, regional plans are steering in character, but their provision needs to be considered by both lower-tier regional plans and by urban plans So technically, there is a hierarchy ranging from higher-order regional plans to lower-order urban plans Note, however, that legally, urban plans have the same force. In other words, higher-order urban plans, such as the General Urban Plan, have the same legal force as lower-order urban plans, such as the Zonal Urban Plan. In other words, the technical hierarchy is not reflected into a symmetrical legal hierarchy
Identify (if any) which acts at national level relate to spatial planning matters	There are two main planning documents that shape spatial development at the national level: the National Territorial Development Strategy (Strategia de dezvoltare teritorială a României/SDTR) and the National Spatial Plan (planul national de amenajare a teritoriului/PATN) Hence, the National Territorial Development Strategy substantiates all sustainable development strategies, programmes, and projects. The National Spatial Plan is the synthesis of all strategic sectoral programmes for the medium and long term. It comes in sections (e.g., transport networks, water, protected areas, the settlement network, natural hazards, tourism, and, still pending approval, education, and rural areas) and its provisions are mandatory for lower-tier (i.e., regional) spatial plans
Identification of specific solutions	In Romania, there are no specific solutions that would allow investments to be implemented independently from planning documents
Identify how environmental and nature conservation sections are included in spatial planning acts at local level	The principles of environmental protection and nature conservation are defined under Romanian legislation Strategies of urban development and urban plans shall consider all the contents of those acts. Failure to take them into account creates the possibility of annulment of the urban plans or strategies Urban plans adapt the restrictions enforced by law to local conditions
Problems with spatial plans at local level	Typical problems faced by urban plans are: • Vulnerability of the proposed zoning and building regulations in the Court of Law. Note that the courts expect both a technical and a legal substantiation of such regulations • Excessive fragmentation of different land use areas, • Lack of clarity or precision of proposed zoning and building regulations, etc.
Defining the role of the courts in relation to planning acts at local level	In Romania, a General Urban Plan can introduce restrictions or easements that can indirectly impact property value Such restrictions can be attacked in a Court of Law and can lead to changes of the original zoning rules in favour of the plaintiff
Frequency of changes in the spatial planning system	The 2001 Regional and Urban Planning Act remains in force in its basic, unchanged scope However, both this Act and other related acts have been amended multiple times Also, more recently, Law 350 of 2001, together with other related acts, is part of a legislative project aimed at bringing a unitary legal framework for regional planning, urban planning, and construction. The resulting Code is entitled the Planning and Building Code (*Codul Amenajării Teritoriului, al Urbanismului și al Construcțiilor/CATUC*)

Source Own elaboration

2.11 Spatial Planning in Slovakia

The Slovak law on spatial planning has been in force for a very long time: since 1976. Of course, it has been amended several times (and a new law is being prepared). Undoubtedly, despite the large number of amendments, the fact that the Act has been in force for such a long time makes it difficult to respond adequately to new challenges. Addressing spatial planning issues in other acts is also a hindering factor (Table 2.9).

Local government units in Slovakia are comprised of the council (elected by universal suffrage) and the mayor. This arrangement is also repeated at the other levels of local government. Spatial plans (which come in two types: general plans and detailed plans) are also a key instrument for spatial planning at the local level. General plans are very elaborate: they contain both text and drawings. In addition to defining zones, they contain legal guidelines for development. These plans also include provisions that provide the basis for integrated development planning. However, integrated development planning is hampered by a comparatively weak strategic planning in Slovakia. Against this background, Slovak spatial plans enacted at the regional level have a very similar content with plans adopted at the local level. While the rule is to obtain building permits based on local spatial plans, there are situations where the content of regional plans will be a sufficient basis for the implementation of investments. However, this applies only to key, strategic investments. There is also a spatial document at the central level. Slovak courts do not review the content of spatial plans. They limit themselves to assessing procedural matters.

2.12 Conclusion

This chapter has attempted to give a broad overview of the spatial planning systems of the countries studied. This goal was achieved firstly by preparing more comprehensive and differentiated tables. However, a careful perusal of these tables leads to the conclusion that the differentiation is not excessive, and its occurrence is justified. It allows for a stronger emphasis on the specifics of each planning system. Short texts are available under each table, aimed at giving the broadest possible overview of the individual systems, while maintaining comparability. The authors believe that this objective has been achieved to a considerable extent. Based on this analysis, we have also identified several important levels where in-depth comparisons and discussion are possible, and indeed, necessary. They will appear in the next chapter. However, it has already become apparent that the common historical context of these countries is reflected both in the dominant spatial planning solutions given and in the widespread intentions for future systemic reforms across all CEE countries.

Table 2.9 A synthetic view of the spatial planning system in Slovakia

Information on local government units in the country (which ones are there?). How many units are there at the lowest level (usually municipalities)? What is the area of the country and the population of the country?	In Slovakia, on the local there are 2927 municipalities (NUTS 5). Total number of municipalities without the urban districts of Bratislava and Košice, including Bratislava and Košice as a whole, there are 2890, out of which 141 have the status of a city, and 8 regions (NUTS 3). The surface area of Slovakia is 49,035 km^2. Population of Slovakia is 5,449,270
Brief description of the local level of public authorities. Brief indication of which authorities participate in spatial planning by drawing up instruments (the local level can be diverse. It can be municipalities, but also counties and cities)	Municipalities have municipal councils and cities have city councils, both elected by citizens. The head of the village is the mayor, and the head of the city is the city mayor. The municipality approves and adopts the land use plan together with the generally binding ordinance of the municipality (VZN), which is declared in the binding part of the land use plan of the municipality. At the regional level, there is the self-governing region with regional council and the chairman ('župan'). The regional council approves and adopts the regional plan. The national government approves the national policy about the spatial planning
Characteristics of legal acts concerning spatial planning	Act 50/1976 Coll. On Spatial Planning and Building Order (the Building Act) as amended. In 2022, two new Acts have been approved Strategic plans on national, regional, and local levels are adopted but are not legally binding. The only legally binding documentation is a spatial plan either on national, regional, local, or zonal levels
Determination of which acts at the local level define the zones (and whether these acts are binding and there is an obligation to enact them)	Act 50/1976 Coll. On Spatial Planning and Building Order (the Building Act) is binding for the Zonal plan, detailed description what the plan should consist of is described in Decree no. 55/2001 Coll.—Decree of the Ministry of the Environment of the Slovak Republic on spatial planning documents and spatial planning documentation The need for a zonal plan elaboration is expressed in the land use plan of a municipality. After the zonal plan is issued, approved, and adopted by the municipality, it becomes a legally binding documentation for the municipality
Identify which acts at the local level set out detailed guidelines for land use—development parameters (and whether these acts are binding and there is an obligation to enact them). Identify what these guidelines are	Decree no. 55/2001 Coll.—Decree of the Ministry of the Environment of the Slovak Republic on spatial planning documents and spatial planning documentation describes in detail what which document should consist of and each developer of the spatial plan must follow these policies, they are mandatory. Policies include description of the textual and graphic part of the Spatial plan (including basic information on the area, surveys and analyses, main aims and strategies, concept, design of the spatial plan including spatial coordination, transportation, nature-protected areas, limits of the development, regulations of the space, etc.) At the bottom of the page, a translation of the detailed content of the land use plan of the municipality is available according to the Decree no. 55/ 2001 Coll.
Determine whether a graphic/map is part of the spatial plan at the local level. If yes—is it a binding part	Land use plans at the local level consist of textual and a graphic part. If approved and adapted in binding act, both text and also graphic could be binding at regional level. At local level, there are text and part of the graphic binding

(continued)

Table 2.9 (continued)

Identify which acts at the local level shape spatial development policy (if different acts, what is more or less in each document)	There are no development policies or acts at the local level
Regional-level characteristics	In Slovakia, there are 8 self-governing regions (NUTS 3). Regional land use plans are approved and adopted by the self-governing region. At the head of a self-governing regional government with the chairman ('župan'), elected by citizens
Identify which acts at regional level set out zoning and land use guidelines (or related elements)	The same as for the municipalities Decree no. 55/2001 Coll.—Decree of the Ministry of the Environment of the Slovak Republic on spatial planning documents and spatial planning documentation describes in detail what which document should consist of and each developer of the spatial plan must follow these guidelines, they are mandatory. Guidelines include detailed description of the textual and graphic part of the spatial plan (including basic information on the area, surveys and analyses, main aims and strategies, concept, design of the spatial plan including spatial coordination, transportation, nature-protected areas, limits of the development, regulations of the space, etc.) either on the regional level or for a zonal plan
Identify which acts at the regional level define development policy in the spatial dimension	There are no special acts. The same act serves for regional as for local/municipal levels. According to this act, there is a regional land use plan developed and approved by the regional council
Brief overview of the relationship between acts at regional level and acts at local level	There are no regional acts. The self-governing region has no right to adopt local acts, only generally binding ordinance
Identify (if any) which acts at national level relate to spatial planning matters	Amendment of the Legal Planning Act, valid from September 1979, established National Planning as Territorial Planning Documentation with adequate legal relevance—the Slovak Spatial Development Perspective (KURS). SSDP is an urban planning tool and is an inseparable component of the Slovak Republic planning system. Its main function is to determine principles and regulations for the development of a settlement system in Slovakia and to specify actual actions for its implementation into practice. The flexibility of the document creates space for further research and consultations with the possibility of further detailed specification and integration of binding and recommended elements. SSDP (KURS) is a long-term strategic document and serves as a synthesiser of all the outputs of the monitoring and knowledge systems undertaken to date. Also, other national, regional, and local conception materials are supportive for territorial planning documentation, the realisation of which has spatial implications. The National Development Plan (NDP) is a tool on the national (NUTS 1) level and is used for application of the regional policy of the SR. The aim is to allow the granting of financial resources from EU structural funds
Identification of specific solutions	In Slovakia, only investments in accordance with the local land use plans receive an official building permit. Under certain circumstances, if a local land use plan is not available, building permit can be issued on the basis of a regional land use plan in case of strategic investments (e.g., highways)

(continued)

Table 2.9 (continued)

Identify how environmental and nature conservation sections are included in spatial planning acts at local level	Act 17/1992 Coll. Environmental Law Act 543/2002 Coll. On Nature and Landscape Protection Act 24/2006 Coll. on Environmental Impact Assessment There are many other laws where there is at least a mention on the issue, but these are the most important. In principle, we have mandatory concerned authorities, whose task is to guard these sections
Problems with spatial plans at local level	Spatial plans as such are relatively functional. The fundamental problem is weak strategic planning (short-term documents) and its bindingness (not binding), the connection between spatial (long term, binding) and strategic planning. Development programme (strategic) is relatively one-sidedly focused and purposefully done for a specific investment, but it does not represent a coherent strategic plan. This is also reflected in the content and objectives of spatial planning, which can result in legal uncertainty of investment, corrupt behaviour and instability in investment
Defining the role of the courts in relation to planning acts at local level	The role of the court is in the control of compliance with the processes, as we do not have an institution that would deal with the control and evaluation of the content of the plans (it is only checked for the procedurality). The newly established office for spatial planning may be a certain hope for the future
Frequency of changes in the spatial planning system	Current Building Act 50/1976 Coll. is from the year 1976 (from former Czechoslovakia) and was updated with 43 Amendments since 1990. In 2022, two new acts have been approved. They will replace the current Building Act from 1976

Source Own elaboration

References

Act CLXXXIX (2011) Act on local governments in Hungary. https://net.jogtar.hu/jogszabaly?docid=a1100189.tv. Accessed 31 Jan 2023

Act CXXXIX (2018) Act on spatial planning of Hungary and some of its priority areas. https://net.jogtar.hu/jogszabaly?docid=A1800139.TV. Accessed 31 Jan 2023

Act LXXIV (2016) Act on the protection of the image of the settlement. https://net.jogtar.hu/jogszabaly?docid=a1600074.tv. Accessed 30 Jan 2023

Act of 14 March (2006) Act No. 183/2006 Coll. on town and country planning and building code (Building Act). Parliament Czech Republic. https://www.mmr.cz/getmedia/9a941cf5-268b-4243-9880-d1b169fb33d6/SZ_angl.pdf. Accessed 25 Jan 2023

Act of 27 March (2003) Act on spatial planning and development. J Laws 741. https://isap.sejm.gov.pl/isap.nsf/download.xsp/WDU20030800717/U/D20030717Lj.pdf. Accessed 10 July 2022

Act of 31 March (2001) Spatial planning. Ministry of Regional Development and Public Works Bulgaria. https://www.mrrb.bg/bg/zakon-za-ustrojstvo-na-teritoriyata-84665/. Accessed 22 Jan 2023

Act of 16 April (2004) Nature Conservation Law. https://isap.sejm.gov.pl/isap.nsf/download.xsp/WDU20040920880/O/D20040880.pdf. Accessed 31 Jan 2023

Act of 27 April (1976) Act on land-use planning and building order (the Building Act), 50/1976 Coll. Available online: https://www.fao.org/faolex/results/details/en/c/LEX-FAOC183199/. Accessed 31 Jan 2023

Act of 27 April (2001) Environmental Protection Law. https://isap.sejm.gov.pl/isap.nsf/download.xsp/WDU20010620627/U/D20010627Lj.pdf. Accessed 31 Jan 2023

Act of 8 May (2008) Development Planning System Law. Saeima of the Republic of Latvia. https://likumi.lv/ta/en/en/id/175748. Accessed 31 Jan 2023

Act of 5 December (1991) Act on environment, 17/1992 Coll. https://faolex.fao.org/docs/pdf/slo45130.pdf. Accessed 31 Jan 2023

Act XXI (1996) Act on spatial development and planning. https://net.jogtar.hu/jogszabaly?docid=99600021.tv. Accessed 31 Jan 2023

Albert C, Fürst C, Ring I et al (2020) Spatial planning in Europe and Central Asia—enhancing the consideration of biodiversity and ecosystem services. Landsc Urban Plan 196:103741. https://doi.org/10.1016/j.landurbplan.2019.103741

Altrock U, Güntner S, Huning S et al (2016) Spatial planning and urban development in the new EU member states. From adjustment to reinvention. Routledge, London

An Act to Implement the Building Code and the Planning Act (2015). https://www.riigiteataja.ee/en/eli/ee/Riigikogu/act/519052022002/consolide. Accessed 27 Jan 2023

Auziņš A, Jürgenson E, Burinskienė M (2020a) Comparative analysis of spatial planning systems and practices: changes and continuity in Baltic countries. In: Hepperle E, Paulsson J, Maliene V et al (eds) Methods and concepts of land management—diversity, changes and new approaches. vdf Hochschulverlag ETH, Zurich, pp 13–22

Auziņš A, Jürgenson E, Burinskiené M (2020b) Comparative analysis of spatial planning systems and practices: changes and continuity in Baltic countries. In: Methods and concepts of land management. vdf Hochschulverlag AG an der ETH Zürich, Zürich, pp 13−22, https://vdf.ch/methods-and-concepts-of-land-management-e-book.html

Building Code (2015). https://www.riigiteataja.ee/en/eli/526022021001/consolide. Accessed 27 Jan 2023

Cotella G (2007) (R)Evolution of Central and Eastern European spatial planning systems: trends towards divergence or uniformity? Cent Eur J Arch Plan 11(2):11–19

Cotella G, Janin Rivolin U (2011) Europeanization of spatial planning through discourse and practice in Italy. DISP 47(186):42–53. https://doi.org/10.1080/02513625.2011.10557143

Dąbrowski M, Piskorek K (2018) The development of strategic spatial planning in Central and Eastern Europe: between path dependence, European influence, and domestic politics. Plan Perspect 33:571–589. https://doi.org/10.1080/02665433.2018.1513373

Davoudi S (2007) Territorial cohesion, the European social model, and spatial policy research. In: Faludi A (ed) Territorial cohesion and the European model of society. Lincoln Institute of Land Policy, Cambridge, pp 81–103

Decision No. 525 of June 27 (1996) The approval of the General Urban Planning Regulation. https://legislatie.just.ro/Public/DetaliiDocumentAfis/9499. Accessed 26 Jan 2023

Decree No. 55/2001 Coll. (2001) Decree on land-use planning materials and land-use planning documentation. Ministry of the Environment of the Slovak Republic

Decree No. 218/2009 (X. 6.) (2009) Government decree on the content requirements of the spatial development concept, the spatial development program and the territorial planning plan, as well as the detailed rules for their matching, elaboration, negotiation, acceptance and publication. https://net.jogtar.hu/jogszabaly?docid=a0900218.kor. Accessed 31 Jan 2023

Decree No. 253 (1997) Government on national urban development and building requirements, XII. 20

Decree 419/2021 (VII. 15.) (2021) Government decree on the content of settlement plans, the procedure for their preparation and acceptance, as well as on specific legal institutions for settlement planning. https://net.jogtar.hu/jogszabaly?docid=a2100419.kor. Accessed 31 Jan 2023

Environmental Protection Law (1992) Czech National Council Act No. 114/92 on the protection of nature and the landscape. https://www.ecolex.org/details/legislation/czech-national-council-act-no-11492-on-the-protection-of-nature-and-the-landscape-lex-faoc004729/. Accessed 27 Jan 2023

Estonia 2035. https://valitsus.ee/en/media/3926/download. Accessed 30 Jan 2023

Feranec J, Soukup T, Taff GN et al (2017) Overview of changes in land use and land cover in Eastern Europe. In: Gutman G, Radeloff V (eds) Land-cover and land-use changes in Eastern Europe after the collapse of the Soviet Union in 1991. Springer, Cham. https://doi.org/10.1007/978-3-319-42638-9_2

General Regulations for the Planning, Use and Building of the Territory (2013) Cabinet of ministers. https://likumi.lv/ta/en/en/id/256866. Accessed 23 Jan 2023

Kowalewski A, Nowak MJ (2018) Chaos and law. Conditions, processes, effects, recommendations (in Polish). Studia KPZK PAN, CLXXXII, Warsaw

Law No. 50/1991 of 29 July (1991) Law regarding the authorization of construction works. Parliament of Romania. https://www.isc.gov.ro/files/2020/Legislatie/Lege%2050%201991%28r2%29.pdf. Accessed 31 Jan 2023

Law No. 350/2001 of July 6 (2001) Law on regional planning and urbanism. Parliament of Romania. https://isc.gov.ro/files/2019/Legislatie/Lege%20350%202001.pdf. Accessed 26 Jan 2023

Law of the Republic of Lithuania on the Protected Areas (2001) Seimas of the Republic of Lithuania, No IX-628. http://www.seniejitrakai.lt/law-of-the-republic-of-lithuania/. Accessed 25 Jan 2023

Law on Environmental Impact Assessment (1998) Parliament of the Republic of Latvia. https://likumi.lv/ta/en/id/51522-on-environmental-impact-assessment#_ftn1&pd=1. Accessed 31 Jan 2023

Law on Territorial Planning of the Republic of Lithuania (1995) Seimas of the Republic of Lithuania, No I-1120. https://e-seimas.lrs.lt/portal/legalAct/lt/TAD/dde75b13095011e78dacb175b73de379?jfwid=wny8rfncr. Accessed 25 Jan 2023

Law on the Protection of Immovable Cultural Heritage (1994) Seimas of the Republic of Lithuania, No I-733. https://e-seimas.lrs.lt/portal/legalAct/lt/TAD/09dce1b073ab11e494a19da07b82f985?jfwid=. Accessed 25 Jan 2023

Local Government Organisation Act. Passed 02.06.1993. https://www.riigiteataja.ee/en/eli/530082021001/consolide

Markowski T, Nowak M (2022) Koncepcja Prawa Do Miasta w Polskim Systemie Planowania Przestrzennego. Wymiar Ekonomiczny, Prawny i Polityczny. In: Prawo do Miasta a Wyzwania Polityki Miejskiej w Polsce. Scholar, Warsaw, pp 9–31

National Development Plan of Latvia, for 2021–2027 (2020) Decision of the Saeima of the Republic of Latvia No. 418/Lm13. Cross-Sectoral Coordination Center Riga. https://www.pkc.gov.lv/sites/default/files/inline-files/NAP2027__ENG.pdf. Accessed 31 Jan 2023

National Long-Term Thematic Plan for Public Infrastructure Development in Coastal Area (2016) Spatial planning department of Ministry of Environmental Protection and Regional Development, Latvia. https://unece.org/fileadmin/DAM/env/documents/2017/EIA/CASE_STUDY_3_LV_ThematicPlanForCoastalArea.pdf. Accessed 31 Jan 2023

National Spatial Plan "Estonia 2030+" (2012) The National Spatial Plan, order No 368 of the Government, 30 Aug 2012, Tallinn. https://eesti2030.files.wordpress.com/2014/02/estonia-2030.pdf. Accessed 29 Jan 2023

Nowak M, Petrisor A, Mitrea A et al (2022) The role of spatial plans adopted at the local level in the spatial planning systems of Central and Eastern European countries. Land 11(9):1599. https://doi.org/10.3390/land11091599

Pallagst KM, Mercier G (2007) Urban and regional planning in Central and Eastern European countries—from EU requirements to innovative practices. In: Stanilov K (ed) The post-socialist city. The geojournal library, vol 92. Springer, Dordrecht. https://doi.org/10.1007/978-1-4020-6053-3_23

Parliament of Estonia (Riigikogu). General part of Environmental Code Act. Passed 16.02.2011. https://www.riigiteataja.ee/en/eli/517062022003/consolide

Planning Act (2015). https://www.riigiteataja.ee/en/eli/ee/Riigikogu/act/525102022002/consolide. Accessed 27 Jan 2023

Regional Development Act (2008). https://www.mrrb.bg/en/regional-development-act/. Accessed 22 Jan 2023

Regional Development Law (2002) The parliament of the Republic of Latvia. https://likumi.lv/ta/en/en/id/61002#_ftn1&pd=1. Accessed 31 Jan 2023

Regulations of the Territory Development Planning Information System (2014) Cabinet of ministers. https://likumi.lv/ta/id/267923-teritorijas-attistibas-planosanas-informacijas-sistemas-noteikumi. Accessed 31 Jan 2023

Regulations on Local Government Spatial Development Planning Documents (2015) Cabinet of ministers. https://likumi.lv/ta/id/269842-noteikumi-par-pasvaldibu-teritorijas-attistibas-planosanas-dokumentiem. Accessed 31 Jan 2023

Republic of Estonia Ministry of Finance State (2022) Local governments, spatial planning. https://www.fin.ee/en/state-local-governments-spacial-planning/spatial-planning. Accessed 30 Jan 2023

Siedziako M (2022) A Polish invention or a copy of the Soviet model? Electoral practices during parliamentary elections in Poland under the communist rule (1944–1980). Contemp Eur Hist 1–19. https://doi.org/10.1017/S0960777321000783

Spatial Development Planning Law (2011) The Parliament of the Republic of Latvia. https://faolex.fao.org/docs/pdf/lat174443.pdf. Accessed 31 Jan 2023

Stead D, Nadin V (2010) Shifts in territorial governance and the Europeanization of spatial planning in Central and Eastern Europe. In: Adams N, Cotella G, Nunes R (eds) Territorial development, cohesion and spatial planning. Routledge, London, pp 182–205

Sustainable Development Strategy of Latvia until 2030 (2010) Saeima of the Republic of Latvia. https://www.pkc.gov.lv/sites/default/files/inline-files/LIAS_2030_en_1.pdf. Accessed 31 Jan 2023

Taimsaare E, Jürgenson E (2022) Estonia. In: Halleux J-M, Halleux J-M, Hendricks A, Nordahl B, Hendricks A, Nordahl B, Maliene V, Maliene V (eds) Public value capture of increasing property values across Europe. vdf Hochschulverlag AG an der ETH Zürich, pp 83–88. https://doi.org/10.3218/4147-7

Chapter 3
A Comparison of Spatial Planning and Development Instruments (Related to Spatial Objectives) in the Different Countries

Abstract The chapter identifies key similarities and differences in the spatial planning systems studied. The results obtained are related to the previous discussion. Detailed spatial planning arrangements in the selected countries are then extracted. These solutions are presented in the form of case studies. The case studies were extracted using the following criteria: (1) the originality of the specific detailed solution, (2) the weakness/ineffectiveness of the specific detailed solution, and (3) the connection of the specific detailed solution with the sectoral sphere, in particular with the environmental sphere. The case studies were described according to similar schemes. The presentation of the case studies gives a broader view of the diverse possibilities in spatial planning systems. It also illustrates the possible consequences of such solutions and highlights related problems.

Keywords Spatial planning law · Specific planning instruments · CEE countries · Comparison of spatial planning systems

3.1 Introduction

Following the understanding of the basic concepts (cf. Chap. 1), and the detailed depiction of planning systems across the selected CEE countries (cf. Chap. 2), the next step is to produce an in-depth comparison of their salient features. This comparative analysis will take the form of case studies. Given the variations across national planning system, a commentary is necessary in each case. It identifies particularly relevant themes from a perspective of future scientific discussions.

A separate part of this chapter is comprised of a more in-depth presentation of interesting, specific solutions in selected countries. This material also provides a reference point for future academic debates.

M. J. Nowak et al., *Spatial Planning Systems in Central and Eastern European Countries*, SpringerBriefs in Geography, https://doi.org/10.1007/978-3-031-42722-0_3

3.2 Comparing the Features of National Spatial Planning Systems

To achieve the objectives set out at the beginning of the book, the following issues were subjected to a broader comparative analysis:

1. Legal coverage of spatial planning issues,
2. Spatial planning at the local level: framing spatial plans and the influence of public authorities at different levels,
3. The relationship between spatial planning and development policy at the local level,
4. The role of courts in spatial planning,
5. Specifics of spatial planning at the regional level,
6. Specifics of spatial planning at the national level,
7. Key problems of national spatial planning systems,
8. Specific solutions.

1. The first issue that requires wider critical comparisons is the way in which spatial planning issues **are covered in acts (laws) at the national level**. Although seldomly addressed within the literature, the legal framing of planning issues is important. There are several similarities in this respect across CEE countries. Firstly, each country has enacted a law that contains basic provisions on spatial planning. Secondly, spatial planning issues spill over into other laws and regulations. Consequently, there is usually little coherence across the normative framework. The problem stems from the fact that laws that approach spatial planning from a sectoral perspective often do so in different ways, while using different terminology. In addition, one finds it difficult to separate strict construction law from development parameters that appear in spatial plans. And thirdly, in many CEE countries, the 'main' spatial planning law has been in force for quite a long time. Hence, attempts are made to replace regulations that are inadequate to meet modern challenges (or regulations that are ineffective). However, there is a strong preference in this respect for making minor adjustments to laws. Consequently, the number of such revisions—relating especially to spatial planning laws—is very high.

The statutory treatment of spatial planning issues across the surveyed countries is therefore similar in many respects. Particularly interesting is the 'breakdown' of the spatial planning topic into different legal acts (with the related terminological problems) and the very large number of amendments to the 'main' spatial planning laws. Thus, the approach of the legislator at the national level is very cautious, requiring numerous corrections and additions to planning law. One gets the impression that there are no elaborate concepts concerning the role of law in spatial planning [I didn't quite get this sentence]. Matters are complicated further by a high degree of terminological diversity, which appears even at the stage of diagnosing spatial planning problems.

2. In all countries studied, plans are the key instrument for spatial planning at the **local level**. However, some discrepancies require attention. They relate to the number of spatial plans at the local level and the detailed scope of their content. Across CEE countries, spatial plans generally fit into the following three categories:

- General plans: They define zones and the basic elements of land use. General plans are usually compulsory, being enacted for the entire area of the municipality. They are commonly adopted in CEE countries.
- Detailed plans: They define detailed development parameters and are adopted for specific or problematic areas within municipalities. In practice, however, in most countries they are not commonly adopted.
- Specific plans: They are usually dedicated to particularly significant investments.

Obviously, all these spatial plans are implemented differently in their respective planning systems. Often not all categories of spatial plans exist. Sometimes, the detailed features of spatial plans are more strongly differentiated. The binding power of individual spatial plans also varies. At least in some of the countries surveyed, one can observe a tendency of public authorities trying to ignore the system legislation. The best examples of this are Bulgaria (where general spatial plans are often not enacted) and Poland (where spatial plans are enacted to a limited extent).

In local spatial plans, drawings play an important role, as they form an integral part of the norms that are binding for investors. Most often, they define the demarcation lines of areas with different land uses. Against this background, major discrepancies between the contents of the spatial plans appear in some of the surveyed countries, which should be regarded as a serious systemic problem. They include inconsistencies between individual categories of spatial plans (e.g., Czech Republic), as well as differences between the textual and graphical parts of spatial plans (Poland).

It is possible to define a general typology of local spatial plans in CEE countries, albeit with the caveat that the presentations of this typology will vary across planning systems:

- Firstly, local spatial plans usually contain binding provisions. Hence, they provide excellent grounds for discussing the role of law in spatial planning.
- Secondly, they generally contain drawings. As mentioned earlier, graphics form an integral part of their provisions. This is another interesting topic for discussion, especially from a normative perspective.
- Thirdly, terminological differences produce some confusion when comparing planning systems, but they also make for a rewarding lexical study. Take, for example, the meaning of the term 'zoning'. It reflects different planning cultures, interventions, and modes of action also considered in other previous studies, where special planning takes place 'in the country, in the national context and language' (Dühr et al. 2010).

3. A separate issue is **how spatial planning and development policy** are linked at the local level. Here, too, the risk of terminological inconsistency runs high. Therefore, the perspective presented in the first chapter is crucial, because it concerns the relationship between concept and implementation in spatial planning. As indicated earlier, serious barriers exist here, especially in terms of 'translating' the more conceptual postulates into concrete legal norms. On the one hand, there must be a link between concept and implementation, otherwise concepts become impossible to implement. On the other hand, however, there is a risk of 'over-regulating' certain issues. Ultimately, this can contribute to blocking development, while producing an excessive focus on legal and procedural matters.

Across CEE countries, two types of acts can usually be distinguished:

- Local development strategies,
- Documents that can be labelled 'spatial development plans'.

Within this classification, the specific planning practices across CEE countries will vary. Development strategies are basic strategic documents that usually define directions for development at the local level. These directions may or may not include a spatial context. In turn, documents referred to as 'spatial development plans' may define (non-legally binding) priorities from a spatial planning perspective. Regardless of the specific formula adopted for individual documents, the problem of the relationship between the conceptual and implementation spheres remains topical. Attempts appear in some countries, such as Estonia, Bulgaria, and Latvia, to develop 'integrated spatial development plans and strategies'. Obviously, integrated planning has a broad dimension. However, there are national systems, such as the Czech system, where strategic issues are not addressed at the local level. Nevertheless, an interesting case is that of Lithuania, where spatial development priorities are part of spatial plans (not spatial development plans). A similar solution appears in the Hungarian system.

The issue at stake here is very important from a research perspective. There is no single easy answer for how to reconcile the conceptual sphere and the implementation sphere, i.e., to translate spatial planning goals and practices into a viable normative framework. The tension here is perennial: On the one hand, workable links obviously need to exist between the two spheres. On the other, these links cannot be all too detailed. Furthermore, proper legislation is not the key to every spatial planning problem. For example, the indication in the Polish legislation that the spatial objectives included in the development strategies must be included in the spatial planning instruments will not be sufficient for their straightforward implementation (there is no guarantee that these objectives will be included in the spatial planning instruments in a correct way, i.e., according to the intentions of the strategy's authors). Undoubtedly, this problem must be dealt with in detail in each national planning system. However, some analogies do exist:

- The array of documents in which conceptual issues related to spatial planning can be included is somewhat similar in all CEE countries.
- Problems related to the relationship between these documents are also quite similar, even to the point that there are no systemic answers on how to integrate them into a coherent sequence that spans from the general to the particular.

Against this background, a tentative solution is to formulate specific guidelines in strategic documents (or spatial development plans). A more comprehensive solution is to draft strategic documents in a way that ensures the legal interpretation of spatial plans [Please check whether I got this right]. This involves more of a soft impact. As legally binding acts, spatial plans are subject to varying interpretations in practice. It is often the case that investors contest the solutions they contain. The provisions of both development strategies and spatial development plans should contain a justification of future solutions to facilitate further action by public authorities.

4. The **role of courts** in spatial planning is highly differentiated and determined not only by the planning tradition, but also by the specificity of legal solutions found in each system. From an international perspective, the sheer extent of court interference in spatial planning acts varies widely. A preliminary point of reference for such an analysis is the case of Poland. Here, there are special administrative courts whose task is to verify administrative acts and decisions issued by public administration bodies. These courts are also very active in dealing with spatial planning instruments. It is possible for property owners to challenge a specific spatial plan before a court. In such a situation, the court can verify not only procedural issues, but also the legitimacy of individual substantive planning decisions contained within the plans. Consequently, courts strongly influence the direction of local spatial policy.

The example of Poland is probably one of the most extreme in this respect. Nevertheless, it provides a basis for an in-depth review of this issue as well. The role of courts in Bulgaria, Estonia, Romania, and the Czech Republic is somewhat similar. Here too, courts make not only a procedural but also a substantive assessment of spatial plans. The situation is a little more complicated in Latvia. General plans are reviewed by the Constitutional Court, while detailed plans are reviewed by ordinary administrative courts. The procedure for challenging planning decisions in Lithuania is also extensive, as it involves prior participation of the complainant in public participation procedures (concerning spatial planning in the area). In Hungary, the role of the courts is reduced to verifying the financial consequences of the enactment of spatial plans.

Classifying national planning systems by role of courts is a major challenge. Nevertheless, such an exercise seems important and much needed in current research. Hence, the first step would consist of distinguishing between systems in which courts have an influence over spatial planning instruments and systems in which they have none. Then, planning systems in which there is such an influence could be further classified into systems in which judicial interference is merely procedural and systems where it is substantive. The classification of systems in this latter group is more problematic, with most CEE countries belonging to this category. Possible classification criteria include:

- The extent of judicial interference,
- The degree of procedural complexity,
- The link between contesting spatial planning instruments and the involvement of individual actors in public participation procedures.

5. Another interesting topic is the profile of **spatial planning at the regional level**. There is considerable variation in the size of regions across CEE countries. As a rule, spatial planning acts at the regional level are not legally binding for investors. Examples include Bulgaria, Lithuania, and Poland. In practice, their content is poorly translated at the local level. Again, Poland is a good example here. The provisions of the voivodeship spatial development plan are applied in local planning acts only to a limited extent. In fact, only the location of public interest investments (already specified in the voivodeship spatial plans) will be relevant in this respect.

The problem of transferring regional guidelines to the local level also exists in the Czech Republic, despite the existence of separate planning instruments at this level. It is also possible to identify countries, such as Latvia, where there are no acts directly related to spatial issues at the regional level, but where a spatial perspective with guidelines for the local level is included in development strategies. A strong relationship between the regional and local levels is noticeable in Estonia. Here, it is possible to amend regional acts from the local level as well. A stronger link is also found in Hungary. Slovakia deserves special mention, where spatial plans at the local level have similar content as spatial plans at the regional level. Additionally, spatial plans at the regional level can, in certain situations, provide a sufficient basis for carrying out investments.

A key dilemma for regional planning in CEE countries is how to translate regional guidelines at the local scale. When drafting regional acts or plans, legislators need to consider the viability of the proposed guidelines at the local level. Then, they need to figure out how to reach an agreement between the regional and the local levels. Ideally, communication and streamlining should go both ways, i.e., top-down and bottom-up.

Obviously, spatial planning is more conceptual (strategic) at the regional scale. Thus, it is not necessarily desirable or necessary in every case to introduce spatial legislation at the regional level. This does not change the fact that there is a certain lack of effective translation of regional acts into local practice in many countries, which can become a serious problem, not only for CEE countries, but also for western European countries. In most countries, there are compatibility requirements between certain regional and local acts. However, these requirements often remain vague. They are open to varying and sometimes conflicting interpretations. This vagueness hampers effective translation at the local level. There are, nevertheless, some solutions to this conundrum:

- Deepening the linkage between regional and local provisions: This includes extending planning integration (including special acts dealing with integrated planning), but also the possibility of influencing regional provisions from the local level.
- A limited, but noticeable, binding force for regional provisions (Slovakia). This consideration may apply to specific categories of public investment.

A separate issue is the possibility of creating a solid axiological base for provisions at the regional scale, which would then translate into legally binding provisions at the local level. This is essentially a legal substantiation at the regional level, which can aid rulings in systems where courts play an important role. It can lead to the recognition of spatial order or the public interest of spatial planning activities as key values, which can, in turn, protect space from conflicts.

6. When it comes to the specifics of spatial planning at the national level, there are considerable variations. They start with Hungary, where at the national level there is an act on land use plan that defines spatial development priorities and frames spatial plans at the lower levels, and continue with the Czech Republic, where spatial development policy is addressing spatial arrangements and functional uses of the territory. Finally, there is Lithuania, where the comprehensive state plan establishes the national spatial structure, mandatory provisions, and universally binding requirements for the use of the country's territory.

Some additional examples also come in handy. Hence, in Romania, all sustainable (territorial) development strategies, policies, and programmes are based on the Romanian Territorial Development Strategy. Spatial plans are the preferred instruments for ensuring their application.

In Estonia, in addition to the national strategy, which includes strategic spatial issues, and the national spatial plan, which tackles the most general spatial matters and provides guidelines for county plans, a maritime spatial plan and a nationally designated spatial plan, which anticipates the possibility of locating key national investments, are also drafted.

In Latvia, where there is no separate act at the national level governing land use, spatial policy objectives are included in the development strategy, while maritime issues are addressed within a maritime spatial plan. Furthermore, there are other two additional acts that include spatial development policies, one for the shorter period, guiding implementation of the main spatial perspective on land, and the other for the special (coastal) area.

There are also countries where currently there is no separate act dedicated to spatial matters at the national level. Examples include Bulgaria, Poland, and Slovakia. In Poland, issues related to spatial planning are included in national development documents, but only to a limited extent. In Bulgaria, there **are two documents at the national level**: The National Concept for Spatial Development (2013–2025) and the National Strategy for Regional Development (2012–2022), with the national concept being the instrument that aims to coordinate territorial development processes and sectoral policies. Both documents are governed by the Regional Development Act and both are strategic (not regulatory) in nature.

7. The final issue is related to **problem identification**. Firstly, there is a weak link between spatial planning and financial instruments. On the one hand, such a weak link translates into compensation claims by property owners. On the other hand, it also brings about difficulties in adjusting, overhauling, and expanding technical and road infrastructure. Secondly, there is conceptual confusion in many countries. Obviously, spatial planning encompasses diverse perspectives. Consequently, the terminology used varies greatly. Nevertheless, terminological differences significantly hamper the achievement of planning objectives, especially when they are not properly defined. Poorly defined planning objectives and terminological obscurity thus lead to the misalignment of the different sectoral perspectives in spatial planning.
8. Some of the **spatial planning systems studied** lack an overall development concept, as well as a translation of this concept into individual instruments. Consequently, it is possible to identify cases where a link is missing between the content of a specific local-level act (e.g., a detailed plan) and the broader planning concept. Of course, such an explicit link is not always necessary. A problem arises, however, when solutions at the local level differ significantly from the general assumptions at the higher levels. Hence, there is often a barrier when translating the content of one spatial planning instrument into another. This even applies to different types of spatial plans. In addition, frequent changes to spatial planning regulations are also identified as a problem. They exacerbate such incongruences even more.

Very similar spatial planning problems and challenges were diagnosed across CEE countries. This provides for a sound platform dedicated to future scientific discussions. In each national planning system, the roots of the problems were diagnosed in detail. Overall, it seems that most national systems do not cope well with the task of ensuring coherence between the different spatial planning instruments, with both terminological and methodological discrepancies appearing routinely.

3.3 Case Studies of Interesting Developments in the Spatial Planning Systems of the Countries Studied

The next stage of the analysis comes with case studies. Authors from the selected countries will present case studies from their systems. These concern solutions that the team of authors found particularly interesting and relevant during the general analysis (in Chap. 2). Cases are diverse and therefore receive slightly different presentations. Nevertheless, all case studies adhere to the same general guidelines.

3.3.1 Bulgaria

Since the 1960s, Bulgaria has been a regional destination for cultural, seaside, mineral springs, and mountain tourism. Before the 1990s, the country was usually chosen by tourists from the Eastern bloc. After the fall of the communist regime, it gradually became a destination for a wider range of international tourists (Vodenska 1992). As a result, certain smaller towns on the Black Sea coast or in some of the various mountain ranges have experienced extreme investment pressures to urbanise virgin landscape territories, with new hotels and apartment buildings rising sharply in numbers. The government and municipal authorities abandoned long-term urban planning exercises, thereby allowing investment pressure in real estate to easily materialise. Meanwhile, the necessary urban infrastructure—streets, water, sewage, electricity, urban parks—were unable to keep pace. Environmental protection restrictions were rarely respected: on the contrary, they sometimes became the last resort for the civil society protesting against speculative urbanisation.

In Bulgaria, the law requires each municipality to have a General Spatial Plan (GSP), which provides a general vision about how the region should develop. The Ministry of Regional Development is providing the financial support for drafting these planning instruments. In practice, however, many General Spatial Plans are still missing, with the deadline for drafting them being continuously postponed over the past decade. At the time of writing, more than 50 (of 265) municipalities still do not have a GSP enacted (Dimitrov 2022). In the following paragraphs, we describe two examples: the municipalities of Bansko and Tsarevo.

First Case Study: The General Spatial Plan for Bansko

Bansko is a town located on the fringe of the Pirin Mountains, 160 km south from the capital Sofia. It became famous mostly through winter sports and ski slopes, with Pirin being one of the three National parks in Bulgaria and a Natura 2000 site. The municipality contains two towns, Bansko and Dobrinishte, as well as six rural settlements. The population has remained relatively stable over the past 30 years—around 13,000 people, with more than 70% living in the town of Bansko.

The previous General Spatial Plan for the municipality dated from 1998, replacing regulations within the law of 1978. With an obsolete Plan, the development of the municipal territory became subject to some piecemeal decisions through Detailed Spatial Plans. In 2016, a new General Spatial Plan was approved by the Municipal Council, but this has come after some 15 years of ad hoc real estate development. As an example, between 2004 and 2007 the compact town of Bansko has enlarged its built-up area towards the south towards the ski slopes with around 110 ha, i.e., about one-fourth of its initial built-up area, while the number of beds has risen fourfold—from around 5500 to about 20,000. The ski infrastructure, located in the mountains above the town, was given to a concessionaire company in 2001 and since then the company has been trying to enlarge the ski zone. This further development has been initiated through proposed changes in the National Park Pirin Management Plan, which could potentially allow development in 48% of the territory of the National

Park (Dogramadjieva 2018). Green NGOs have been campaigning against these changes because of concerns that the ecological capacity of the region cannot bear the impact of the potential new developments. The steep rise of tourism in the region has been criticised for lack of necessary municipal infrastructure, creating a non-diversified tourist product that has a peak season, as well as low-quality labour conditions, against a background of global climate change that reduces snow quantities across Europe.

In the absence of a General Spatial plan for the Municipality of Bansko, all the pressure for new real estate developments has been focused in diverting a future general spatial vision through alterations of the National Park Pirin Management Plan and through drafting new Detailed Spatial plans for the zones where new hotels, aparthotels, and ski infrastructure (new lifts, etc.) were planned by the concessionaire company.

Second Case Study: The General Spatial Plan for Tsarevo

The other example of a municipality that works without a general vision for its territory is the municipality of Tsarevo, located on the southern Bulgarian Black Sea coast. The municipality is comprised of two major towns, Tsarevo and Ahtopol, and 11 rural settlements. From 2003 on, the municipality has been trying to draft a General Spatial Plan (GSP), but some 19 years later it has still not succeeded. At least two projects from different urban planners have been produced and both have been proposing to massively urbanise the coastal areas between the settlements. Nevertheless, both procedures have been abandoned and the municipality will start a new plan from scratch.

More specifically, the latest General Spatial Plan (abandoned in 2022) proposed to increase low-density zoning, allowing the development of houses and villas, from 95 to 630 ha for the entire municipality. According to the plan, one small and quite attractive village, Sinemorets, located in the Strandzha Nature Park and on the Veleka River, had its built-up territory enlarged by 40%. The argumentation for this excessive urbanisation was not grounded on reliable data or profound research at all. Urbanisation proposals within the GSP were most likely following the structure of properties and the speculative intentions of their owners, neglecting the long-term effects on the environment (Naydenov 2022). The territory of the municipality of Tsarevo overlaps with Strandzha Nature Park, one of the 11 Bulgarian Nature Parks. At the time of writing (2023), the Park still does not have a Management Plan, which would act as a spatial plan that regulates development activities on its territory.

Meanwhile, speculative projects in the municipality have continued through selling public land in the coastal areas and issuing Detailed Spatial Plans (DSPs) without transparency, to build more guest housing and aparthotels. Once a DSP is approved by the Municipal Council, the Chief municipal architect is required (bearing personal legal responsibility) to approve the building permits for the building plots in question, while the civil society and other stakeholders in the process are left without legal instruments for countering these projects. The speculative projects finally pass through the approval of the Regional Inspection for Environment and Waters, whose methodology evaluates each project on its own, without considering the cumulative effect of all projects on the coastline. Thus, a lot of these building permits get to be issued with a low impact on the environment. In the logic of priority, set by the regional public authorities, they bypass at least four national laws that set the environmental preservation in coastal areas as a higher priority than the need for more urban development (Naydenov 2022).

3.3.2 Czech Republic

In the Czech Republic, the spatial planning system is built upon a hierarchical system of plans. At the municipal level, there are general plans (land use plans) and detailed plans (regulatory plans).

Case Study: Regulatory Plans

The principles of land use plans in the current Czech legislation go back to a system anchored in socialist times, with the Building Code dating from 1976. During those times, spatial planning provisions were not all too detailed. Practical problems were supposed to be solved during the building approval process, via building permits. The permit fixed the position and volume of the building, as well as many other details deemed important for the plot and its neighbours.

Against this background, the new Building Code of 2006 introduced a new planning instrument—the regulatory plan. However, the regulatory plan did not alter the existing concepts of land use plans and building permits. In practice, developers continued to obtain building permits based solely on the land use plan. The idea to use a regulatory plan as a document for issuing bulk building permits for an entire block was rarely applied. Just a few pioneering municipalities tested this new instrument in specific areas, such as historical centres and greenfield developments. It is also interesting to note that the legislator even proposed a subtype of the regulatory plan which does not provide any right to build. The point of this specific type of regulatory plan remains a mystery.

Indeed, there was strong hope that after adoption of the Building Code in 2006, the regulatory plans will increasingly complement existing spatial planning instruments, but this never happened. Regulatory plans remained sidelined and were not welcomed by politicians, due to the extra effort that went into drafting them that seemingly failed to pay off. The introduction of regulatory plans, without changing the role of land use plans and without providing them with a specific role within the Czech spatial planning system, has led to the opposite outcome than the one initially envisaged. Instead of simplification and clarity, complications emerged, with planning provisions facing even more barriers to efficient implementation. Land use plans remained fully binding documents, combining a general vision and an overall concept for development with rather detailed regulations for specific plots, and the regulatory plan brought almost no added value to this approach.

Nevertheless, never-giving-up experts and politicians tried to find a way out, by adopting 27 building code amendments within 13 years. Unfortunately, their main result was disappointing: The number of days needed for building permits increased from 150 to 247 in the Czech Republic. Finally, a completely new Building Code was approved in 2021 (still not in force in 2023). Unfortunately, the inconsistency of general and detailed plans escaped the attention of the legislator yet again.

To sum up, it has become poignantly obvious that remaining captive in the past will not provide viable answers to future challenges. Hence, in the spirit of the Aarhus Charter, effective public participation in planning requires ample time dedicated to participation, negotiation, and even disputes. It therefore seems smart to split that process into pieces and sort the issues to be solved, viz. to have a general discussion about the general plan and later a detailed discussion about the detailed plan. Without such preparations, it becomes almost impossible to approve the land use plan in a municipality within a reasonable time. And this, unfortunately, is the current situation in the Czech Republic.

3.3.3 Estonia

The first legal act that regulated planning activity in Estonia after regaining independence, the Planning and Building Act, was adopted in 1995. Separate laws for the planning and building sector were adopted in 2002. The current Planning Act has been valid since 2015 and provides the foundations for organising planning activities. Several changes in the sectoral regulations notwithstanding, some principles have remained valid since this first regulation. One is that the Estonian planning system has four levels (Auzinš et al. 2020, 2022). Hence, the national spatial plan and the county spatial plan fall under state responsibility, while the comprehensive and the detailed plan falls under municipal responsibility. The comprehensive spatial plan defines the principles and directions at the municipality level. The purpose of detailed spatial plans is to implement a comprehensive spatial plan and to create an inclusive spatial solution for the planning area. The municipal authority organises spatial planning work (including comprehensive and detailed spatial plans) at the local level. Among other tasks, it must ensure the existence of spatial plans, arrange the drafting or the commissioning of plans, and carry out the procedural operations required in creating a spatial plan.

First Case Study: Outsourcing Plan Preparations in Estonia

Only the administrative body can organise the creation of a spatial plan, as it is a public authority task. In the case of a comprehensive and a detailed spatial plan, the municipality is an essential participant in the planning process, both in the drafting and implementation stages. Nevertheless, creating a detailed spatial plan (working out a planning solution) is a process that is commonly outsourced in Estonia. There is the possibility to hand out some tasks to a private person or entity, based on the management contract. However, within such a contract, the authority may not delegate the creation of the plan or the carrying out of any procedural operations required to create the plan.

In practice, there are different possibilities for drafting a spatial plan:

1. The municipality can delegate the task to one or several public servants who work within it.
2. It can contract a planning consultant (planning company, architectural office).
3. It can also assign the task of drafting or commissioning a detailed spatial plan to an interested party (from a development point of view).
4. The administrative body may have two contracts: one is with a planning consultant (planning company, architectural office), and the other is with an interested party for the financing planning process.

The municipality that organises spatial planning work ensures that an authorised planning professional drafts the spatial plan. The Planning Law (Parliament 2015) enacts the term 'spatial planner'. It means a person who possesses a higher education corresponding to a master's degree in geography, architecture or landscape architecture or a person holding a professional specialisation certificate or a person who has been granted the qualification of a spatial environment planner. Hence, there are two ways of becoming a spatial planner: to study geography, architecture, or landscape architecture or to apply for a professional certificate in spatial environmental planning at level seven.

According to the occupation registry, there are 33 persons who have been awarded an occupation certificate for spatial environmental planning at the end of 2022 (Estonian Qualification Authority 2022), in addition to several planners who have completed a master's degree in geography, architecture, or landscape architecture. There are approximately 100 members in the Estonian Planning Association (Eesti Planeerijate Ühing 2022). However, not all members are planners, and active planners have no obligation to join the Association. The possibility to enter the planning profession based on one's educational achievements became possible in 2015, when the current Planning Act was passed, as the occupation standard for a spatial planner was new, and there was a concern that Estonia could not field sufficient spatial planners. Today, this practice has come under scrutiny, and many discussions centre around the sufficiency of academic qualifications for granting entry to the planning profession. It seems that in future only the professional certificate will enable planning practitioners to engage in spatial planning (Konsap 2022).

In Estonia, planning activities are often outsourced to the private sector. This practice is widespread in other fields as well, such as cadastral surveying, for instance. Estonia took over the liberal worldview after regaining independence in 1991 and quickly developed a more aggressive neoliberal stance. It means there is a minimal state that does not intervene to guide entrepreneurship. A lot of services are offered by private entrepreneurs. This direction was chosen immediately after regaining independence, and there was lack of knowledge for leading the state based on a market economy. The need for funding has encouraged the implementation of neoliberalism as well, with the IMF and the World Bank extending credits to Estonia, soon after it regained its independence. Implementing free market principles was the precondition to borrowing money (Harvey 2005).

Second Case Study: Environmental and Natural Provisions Within Spatial Plans in Estonia

The Planning Act determines environmental principles that apply to all planning levels. Planning should contribute to a functional environment that includes both natural and artificial areas. The spatial plan must establish the preconditions for the existence and preservation of a user-friendly and safe living environment and a spatial fabric that reflects the values of the community, as well as the development of aesthetic surroundings. When creating spatial plans, appropriate use of previously used or insufficiently used areas must be promoted wherever possible, preference must be given to environmentally sound solutions that ensure good energy performance, and the production and use of renewable energy must be encouraged. A balance between development and the natural environment must be safeguarded during the planning process. In addition, the Building Code brings in energy performance. The previous land use should always be analysed and appropriately designed, with some areas densified in urban areas, for instance.

When planning human settlements, the built environment and the green open areas must receive equal attention, considering circumstances dictated by the existing environment and the specific location. This is the principle of universality, which states the obligation to consider different documents that protect natural assets. One is the Constitution of the Republic of Estonia, which safeguards natural assets. It affirms that everyone has to preserve the living and natural environment and compensate for the damage they cause to the environment (The Constitution of the Republic of Estonia 1992). The second one is the General Part of the Environmental Code, which sets out measures that must ensure a high level of environmental protection (Parliament 2011). Remedies for the safety of people and the environment must provide adequate protection against environmental disturbance and economic consideration must not be given preference.

The planning process includes a special procedure, called the environmental assessment. Both the Planning Act and Environmental Impact Assessment and Environmental Management System Act regulate this procedure. It is mandatory for all spatial plans, except for the detailed spatial plan, where it is required only if a planned activity has a significant environmental impact (Parliament 2015). Environmental impact means any potential direct or indirect effect of a proposed activity or implementation of a strategic planning document on the environment, human health and well-being, cultural heritage, or property.

Generally, a comprehensive spatial plan consists of several parts. For instance, the recently completed and modern GIS-based (presented as a digital book) comprehensive plan for the city of Tartu consists of parts for land use and construction, spatial development, heritage protection and milieu, nature, engineering networks, and transport. The nature part includes valuable agricultural land and landscapes, green networks and recreation areas, biodiversity, mineral deposits and water bodies

(Tartu Linnavalitsus 2022). All these issues are described and presented graphically on the maps.

The comprehensive plan of Tallinn, capital of Estonia, is designed differently. It is comprised of separate parts for greenery and recreation areas, as well as environmental protection—air, noise, water protection, nature protection, and waste management.

The treatment of environmental issues in a detailed spatial plan depends on the planning task. When there is a need to create a detailed spatial plan, then the fulfilment of some of the tasks is mandatory, and many of the tasks mentioned in the Planning Act are carried out according to these demands.

Hence, environmental provisions are an essential part of spatial plans in Estonia. The principles are unique and apply to all levels (state and municipality). At the same time, environmental issues are regulated with different acts, thereby producing a rather complex system.

3.3.4 Hungary

In Hungary, the Act on development and protection of the built environment and the related governmental decrees clearly set out the procedural framework for spatial planning. However, increasing efforts were made during the past decade to simplify and shorten the process for certain projects. They are mostly large-scale infrastructural projects that are cofinanced by the EU, but lately other types of projects have begun to get exemptions from the rules as well.

First Case Study: The Consequences of Facilitating and Accelerating National Priority Investments

In 2004, the Government passed an act for simplifying large-scale investments, such as motorway constructions, aimed at safeguarding the timeframe for investments cofinanced by the EU. It later became the preferred tool for removing all obstacles facing investments prioritised by the government.

The initiative for reducing the timeframe for large-scale investments started with an act simplifying rules related to expressway construction. Hence, on the 1st of January 2004, the Public Interest and Development of the Expressway Network Act entered into force. It aimed to simplify the tendering procedures for road infrastructure investments, in part by cutting back the rules for public participation.

In the beginning, this legal initiative had sound reasons, as even the European Community encouraged the fastest possible implementation of cofinanced investments. As the EU has a seven-year budget cycle, Member States could spend the financial envelope allocated for a given budget year only for the next two or three years, i.e., the $n + 2$, $n + 3$ rule. Against this background, the European Commission withdraws the part of the budget commitment that has not been used in the given timeframe (Szabó et al. 2016).

The next step followed in 2006, when the National Assembly adopted the Act LIII of 2006 on 'The acceleration and simplification of the implementation of priority investments significant for the national economy'. Its aim was to allow derogations from the general and sectoral procedural rules, based on a separate law, in the case of exceptional large investments, where, for example, the use of EU funding was of particular importance for the rapid completion of permitting and other official procedures. The law authorises the Government to designate in separate decrees priority cases for the national economy, and thus to derogate from the otherwise applicable rules on matters specified in the government decree (INT-1). Unfortunately, since then we have witnessed an increasing number 'exceptional' cases, as the legislator is exempting more and more specific cases (at least partially) from the general and sectoral rules. The consequences have also become poignantly clear, as we can witness the weakening of guarantees that protect clients, environment, landscape, and other procedural safeguards that ensure informed and mature decisions.

The 'Act on the acceleration and simplification of the implementation of priority investments significant for the national economy' (Act LIII of 2006) designates transport infrastructure investments as high priority. It mandates a prior consultation with the authorities defined in the Act 'in order to facilitate the selection of possible routes for transport infrastructure investments', which precedes the environmental permit procedure. The Act shortens of administrative deadlines (e.g., 60 days for the environmental impact assessment, instead of the 105 days under the Environmental Protection Act (INT-2)).

Since 2006, a lot of amendments were made to the Act. Originally, it ensured that public investments of at least HUF 5 billion and creating at least 1000 jobs were prioritised. Among other changes, this threshold has been abolished for projects financed in full or in part from EU or national budget, and for other investments the threshold has been lowered to HUF 90 million and 15 jobs. Furthermore, the possibility to get priority status was widened, the government now has the possibility to lay down the rules of construction, specific building requirements, and spatial planning rules by decree:

- For plots of land involved in construction activities during investments of major national economic importance,
- For plots of land used for cultural and sports facilities of national importance,
- For plots of land in their immediate vicinity, where public interest investments are carried out there, and finally,
- For plots of land used for facilities for national security purposes.

This provides exceptional circumstances for priority investments, as these do not have to apply with the rules of the current spatial plans (INT-1; INT-2). According to Atlatszo.hu, the first Hungarian investigative journalism non-profit organisation, the Act has been amended 40 times since 2010. During these amendments, new concessions have been granted to ensure the smooth implementation of a growing number of priority investments and related utility and transport developments (INT-3). Under the Building Regulation Act, the Minister responsible for building regulation and building authority matters coordinates the activities of the public authorities concerned with facilitating the implementation of construction projects that have a major importance for the national economy (INT-2).

Even without the issues mentioned above, the authority of local municipalities in shaping the built environment was narrowed. In March 2020, the government withdrew the issuance of building permits from municipalities and transferred it to government offices. Since building permits can only be issued by a government office, the powers of local governments have drastically narrowed in terms of what and how can be built on their territory.

The Local Building Regulations determine land use and the type, size, and scale of buildings in each area. Drafting the Local Building Regulations is the responsibility of the local municipality. However, the building permits are issued by a government office. After submitting the application for a building permit, subsequent modifications made to the local spatial plan by the municipality are rendered moot for that submission.

In principle, the municipality can reduce building parameters and tighten the regulations during the amendment of a spatial plan, but the construction law states that if a developer incurs losses to these updated regulations, the municipality must provide compensations. So, for example, the municipality cannot permit the construction of only five-story buildings instead of ten, without incurring penalties (INT-4).

Another tool of the municipality is the right to launch a townscape opinion. After consulting with the council of representatives, the mayor must provide an opinion on the townscape for every building permit application submitted to the government office. A few years ago, this was still a right of veto, so if the local government expressed a negative opinion, the investment could not receive the green light. However, in case an investment is prioritised, the municipality cannot carry out a townscape opinion procedure.

The government regulations define the types of cases that are designated priority investments, with most of them concerning construction, heritage protection, roads, environmental protection, water management, technical safety, real estate (land surveying, land conversion, land registration), mining, fire protection, and forestry. However, the legislator always concludes the list of types of cases with a flexible clause (e.g., procedures for granting î authorisations by public authorities not included in the list, which are necessary for the realisation, putting into service, installation, operation, and management of the investment in question). In practice, therefore, all the authorisation procedures relating to the investment in question are included in the list of priority investments (INT-1).

Initially, priority investments were really carried out in the public interest (metro 4 in Budapest, the Mercedes factory in Kecskemét, or the Pécs European Capital of Culture project, etc.), with relatively few government decrees. In the last decade, however, this number has grown exceptionally, with 1 in 2007, 8 in 2008, 9 in 2009, 40 in 2015, and 69 in 2017, and more than one thousand between 2018 and end of 2021. Nowadays, any project of at least HUF 90 million (EUR 243,000) that creates at least 15 jobs, or meets other similarly easy conditions, can be recognised as priority project by the Government, even if it is private. For the time being, the Hungarian

Government has classified more than 3000 investments as priority projects (INT-5). There is a highly disputed investment in Budapest, in the 14th District, close to Bosnyák Square, the new district centre of Zugló, where a project is carried out on a 7-ha area as priority investment (Government Decree 314/2022 (VIII. 11.)). As such:

- An architectural and engineering design opinion is not required,
- No townscape notification procedure is necessary,
- There is no need to carry out a townscape opinion procedure,
- There is no need for a townscape consultation,
- There is no need to draft and implement a plan for the development of public spaces.

The time limit for issuing the building permits is fifteen days, except for environmental protection and nature conservation procedures, or if no shorter time limit is provided by the law (INT-6). Thus, even though the municipality has repeatedly expressed its opposition to the planned investment on the plots (the area of the investment is divided into 4 plots, so the effects cannot be analysed as a whole), the decree revised the local regulations with new rules, including those on the extent and height of the construction, the organisation of traffic, and removed new municipal powers (INT-7). According to the decree, buildings up to 39 m high can be built in the suburban area of Zugló, where up to 4000 apartments can be developed, with the developer intending to build 900 apartments.

The Regulation also provides additional opportunities, such as no minimum distance between the buildings, the possibility of combining buildings below ground level and at ground level, no limit on the number of buildings and units that can be placed on the plot, and no limitations on the area of each unit. And there is no need for P+R parking plots. There is no obligation to replace trees cut during the construction on the plots concerned and no obligation to provide green roofs.

Second Case Study: Specific Environmental Natural Provisions Within Spatial Plans

In local plans, the regulation must reflect and apply the requirements of the national plan. In Hungary, there is a strong hierarchical system of land use plans, with a detailed national-level plan that gives guidance on spatial development at the local level.

The area covered by the National Ecological Network (EN) includes natural and semi-natural habitats of national importance and the unified and composite system of ecological corridors that connects them. The national plan contains quite basic rules for the EN, such as rules restricting the designation of areas for development, the placement of transport infrastructure, and new surface mines. These regulations indirectly contribute to the protection of biodiversity.[1]

[1] See also the Hungarian Green Infrastructure section of the Biodiversity portal, available at: https://biodiversity.europa.eu/countries/hungary/green-infrastructure [28/08/2023].

For world heritage sites and candidate areas, there are basic rules, such as restricting the placement of transport infrastructure and new surface mines. For landscape protection, the spatial characteristics of the landscape character shall be determined within the framework of the preparatory work part of the county's spatial plan. In addition, the elements, and groups of elements, of the landscape character to be preserved, as well as the local characteristics of the landscape unit and the traditional landscape use shall be determined within the framework of the preparatory study for the settlement planning instrument dedicated to the entire administrative territory of that settlement. Furthermore, there are additional basic rules, such as restricting the placement of transport infrastructure and new surface mines.

For water protection areas, sewage water should be properly managed, besides banning new surface mines. In the area covered by the Balaton Priority Resort Area, in addition to a limited use of chemicals and fertilisers, environmentally friendly or extensive agricultural production may be carried out. However, no new chemical storage and waste treatment facilities may be established, except for composting and transfer stations. The rules for the regulation zones in the national plan are general but they mostly highlight the need to pay specific attention for these issues within local plans. Hence, the local spatial plan needs to consider several environmental issues.

In the settlement structure plan the designation of land use units, the network, and especially new infrastructure elements, are basic prerequisites for the environmental quality of a municipality. Hence, development should focus on less sensitive areas or in areas with lower ecological and environmental value. Setting limits for construction activities, stopping urban sprawl is very important for protecting natural values and for decreasing future environmental load. Against this background, the most important objective of both the regulation plan and the building parameters is to decrease negative externalities and to reduce land use conflicts: zoning, building type, scale, minimum green area, building height, emission limits, etc. (Boromisza 2011).

Different protection areas or protection zones can be defined in the regulation plan, based on different legal rules, such as the Act Nr. LIII of 1995 on Nature protection with restriction, where the prohibition of activities, restrictions on construction, use of materials, land use, or obligation to assess and monitor systematically the environmental processes appear. For example, within local plans, the municipality can identify agricultural areas, where the protection of wetlands, marshes, and reedbeds is necessary, as agricultural areas with limited functions or as semi-natural areas (Tájoló-Terv Kft. 2017). Furthermore, based on OTÉK, protection areas or protection zones can be defined, e.g., along infrastructure lines, thereby ensuring emission limits, if the environmental impact cannot be avoided by other means, preferably on the own territory (Boromisza 2011). Hence, protection areas based on air or water protection zones, waste management, etc., can be defined. In addition, different sensitive area categories can be designated in local plans, such as

- Silent zones, strictly protected areas, or high noise areas,
- Nitrate sensitive areas: groundwater-sensitive, highly sensitive, or particularly sensitive areas.

There are further possibilities to consider environmental issues, especially in the optional, additional chapters. They contain the so-called Landscape planning proposals that focus on internationally, nationally, and locally protected natural values and landscape areas that need protection.

3.3.5 Latvia

There are five planning regions in Latvia. A planning region is a derived public entity with the competence to ensure regional development planning, coordination, and cooperation between municipalities and state institutions. As stated by law, each region drafts spatial development planning documents, both for the long term, i.e., the sustainable development strategy with its spatial perspective, and for the medium term, i.e., the development programme. They all need to be considered at the local level when drawing up municipal spatial development planning documents, i.e., strategies with their spatial perspectives, development programmes, and spatial plans.

Case Study: The Riga Planning Region Strategy Update, an Exercise in Cooperation

Riga Planning Region is Latvia's capital region, located in the central part of the country, around the metropolis of Riga. After the administrative—territorial reform of 2021, the Region has a more compact territory and a unified identity. It includes two cities, Riga and Jurmala, as well as seven neighbouring municipalities. Based on this reform, the Region's Sustainable Development Strategy for 2014–2030 was renewed. It was updated by the Region's Spatial Planning Division specialists, in a cocreation process with municipal planners, under the supervision of a Steering group. Working groups were established for each main theme of the Strategy. Consultative meetings with the bordering planning regions took place about trans-territorial issues. State administration officials and experts participated at the public consultation process as well.

During the elaboration process, the Strategy was updated while integrating the latest EU and state-level policies. In addition, it was developed in accordance with the various municipal strategies. A valuable interplay between regional and local solutions emerged, with planning documents at both levels drafted concurrently. Hence, stances expressed previously in a common dialogue later became regional documents, such as the Action Plan for the Riga Metropolitan Area (Action Plan 2020).

As a policy document, the Strategy embodies the region's spatial perspective, as well as guidelines and recommendations that tackle current and anticipated needs for future policy-making and planning. It outlines the joint activities of municipalities, the state, and the region, as well as the main directions for long-term activities and their landmarks, with specific solutions to achieve them (Riga Planning 2022).

The Strategy is comprised of two main parts: The first part contains development settings, while the second includes the main actions for implementing the Strategy and the vision of the Spatial Perspective (Riga Planning 2022). Given the Strategy's limited validity period, it is supplemented by the 2050 Vision for the region. During public consultations, the 2050 Vision became a separate part of the Strategy, due to its high level of importance.

Sustainable and spatial development are closely interwoven within the Strategy. The regional development vision includes the future look of Riga as a metropolis, regional mobility, as well as urban, coastal, and green open spaces. One of the three strategic goals is devoted to environmentally tolerant lifestyle and places, explained with 'key'words, such as 'resilient region' and 'diverse living environment', and followed by the relevant priorities for the settlement structure, places, the natural environment, and energy.

The Strategy is permeated by the 2050 Vision (Riga Planning 2022). The Vision makes planning recommendations focused on the municipal and state levels. Recommendations are both general and specific, with the latter being specifically addressed to the Riga Metropolitan Area, the coastal area, rural and suburban settlements, etc. They are in graphic and text form and cover thematic directions: the spatial structure of settlements, traffic infrastructure, and the spatial structure of natural areas. The Strategy also includes solutions for strategic development areas and for the cooperation needed for complex development projects and sites, as well as for interregional cooperation areas.

Communication with all participants involved in the development processes is pivotal for the implementation of the Strategy. The main instruments are cooperation platforms between state, regional, and local actors. Cooperation forms may vary depending on the tasks at hand: formal inter-institutional coordination groups, informal public interest groups, short-term or permanent cooperation groups, social networks, etc. In addition, active associations, such as the local government association 'Rīgas Metropole', which includes Riga City and the bordering municipalities, the Alliance of Riga Neighbourhoods, etc. are a contributing factor to successful Strategy implementation. The other main implementation instrument is comprised of projects. The Region has a high (administrative) project capacity, with joint development projects strengthening cooperation platforms and expanding strategic topics, such as renewable energy, sustainable mobility, and tourism.

Having the required competence, the Strategy was updated by Region's own specialists, in cocreation process with municipal planners. This cocreation process produced a common understanding of both the recommendations made and their implications and for spatial plans at the local level.

Note, however, that the Strategy has wider role. It looks over the administrative borders of the Region and recommendations are made not only to the local, but also to state level and to other metropolitan functional areas within the Riga region. Regions lacking administrative power and own financial bases use communicative planning and implementation instruments. Implementation processes involve stakeholders through cooperation platforms and projects. The next foreseen step by the Region is the establishment of a regional development fund, geared towards coordination and implementation of cross-territorial development projects.

3.3.6 Lithuania

The Spatial Planning Act defines territorial planning (spatial planning) as a process, carried out in compliance with the legal requirements aimed at engendering sustainable territorial development. It includes the establishment of land use priorities, measures for environmental protection, public health, heritage protection, creation of residential and manufacturing areas, engineering and social infrastructure systems, as well as creating conditions for regulating employment, the development of activities for the resident population, and reconciliation of public and private interests.

Hence, we can conclude that the solution of environmental issues is inseparable from territorial planning, i.e., the solution of environmental issues in the planning of territories is imperatively enshrined in the law. In addition, when setting planning objectives for a specific territory, it is necessary to consider the public needs, the landscape and biodiversity of a planned territory, its geographical location, geological conditions, the existing urban, engineering, transport and agricultural systems, the interests and rights of managers and users of land and other immovable property and third parties, as well as architectural, **environmental**, public health, nature, and heritage protection requirements, and state and public security and defence needs.

This list is obviously quite long, with the same requirements provided in the implementing legal acts (Rules for the Preparation of Documents of Complex Territorial Planning and others). Currently, 78,777 valid territorial planning documents are registered in Lithuania, and 2272 documents are currently being prepared. Most documents, at least formally, analyse the environmental impact of future proposals or solutions. However, in certain situations, it is a legal regulation that determines possible environmental conflicts or violations. For this study, we will provide two different examples of how important environmental issues are not or cannot be found in compromise solutions that match all interests.

First Case Study: The Klaipėda State Seaport

Klaipėda State Seaport is the most important and biggest Lithuanian transport hub, connecting sea, land, and railway routes from East to West. It is the northernmost ice-free port on the Eastern coast of the Baltic Sea. On the 11th of December 2019, the Government of the Republic of Lithuania has approved the comprehensive plan of Klaipėda State Seaport. This is a territorial planning document of a project of importance to the state.

The strategic directions of the port's development and the spatial development concept are prepared in the comprehensive plan for the next 25 years and concrete solutions for the next 15 years. The approved document is part of the implemented project of special national importance, determining the development of the port territory located in the municipality of Klaipėda city, as well as the internal water area, the territorial sea of Lithuania in the Baltic Sea (outer road of the port) and its related infrastructure. The territorial planning solutions determined by the comprehensive plan create prerequisites for the complex development of the Seaport and related activities in the existing and planned new Seaport development areas, as well as for the development of the communication and engineering infrastructure necessary for these activities.

Concerned about the consequences of the Klaipėda seaport comprehensive plan, the public ('the public' shall mean here one or more natural and/or legal persons, their associations, organisations, or groups.) appealed to the Seimas of the Republic of Lithuania, the Presidency, and the responsible ministries, urging them to pay attention to the associated risks of this development. It is stated that the comprehensive plan of Klaipėda port does not ensure sustainable development and sociological, cultural, environmental, and other public interests. The biggest concern is that the solutions lead to a significant deterioration of the environment and quality of life in the city of Klaipėda, as well as the ecological catastrophe of the Lithuanian coast—the probable erosion of shores and the loss of beaches. The solutions within the Klaipėda seaport comprehensive plan are not compatible with the provisions of sustainable development—to meet the needs of today's society in such a way that they are not met at the expense of future generations, so that to achieve sustainable development, economic, social, and environmental aspects are combined, focusing on the quality of human life. In addition, the summary conclusion of the Strategic Environmental Assessment report states that in all cases of Klaipėda State Seaport development alternatives, the consequences would be direct, long-term, negative, and significant. On the 30th of September 2021, the comprehensive plan of the city of Klaipėda was approved.

A particularly difficult situation arose because two comprehensive plans, the city and the Seaport, were being prepared in roughly the same area at the same time. The comprehensive plan of the city was prepared by the municipality, while the Seaport comprehensive plan, which is an object of state importance, was prepared by the Ministry of Transport and Communications. According to the laws, it is the comprehensive plan of the seaport that has higher authority, and the city, when preparing its plan, must consider the decisions of the comprehensive plan of the Seaport, even if their visions do not coincide. The confrontation continues to this day, because in the opinion of the interested public, the current legal regulation blocks the protection of environmental interests and prioritises economic interests.

Second Case Study: Special Spatial Planning Dedicated to Measures for the Use, Management, and/or Protection of Protected Areas

Boundaries of protected areas are changed during the spatial planning process, when boundary plans are drawn up. However, the conditions or circumstances when there is a need to prepare territorial planning documents and to change the boundaries of the protected area, which have already been established, are not strictly regulated. The State Protected Areas Service indicates that the boundaries of protected areas are determined on a value basis, that is, according to the nature of these protected areas, guided by the principle of reasonableness, with the demarcation or change of boundaries requiring substantiation. The essential principles of determining and drawing the boundaries of protected territories and their zones are established in the rules for the preparation of special plans.

When analysing the planning documents of various protected areas, we could find no clear and specific reasons in their explanatory notes that would have enabled us to judge changes in the boundaries or zones of these protected areas. Furthermore, the specific investigations carried out in that territory are not named, so we were unable to find out neither who nor when they were carried out. Most often, abstract concepts or explanations are used, which can be applied to any change of boundaries and zones of protected areas. In many cases, planning documents do not reflect evidence, scientific research, expert evaluation, or other reasoned explanations as to why certain areas are excluded while others are included. Many protected area planning documents contain abstract, general statements indicating that the boundaries are adjusted in accordance with the following criteria: (a) revision of the previous planning and management provisions, changing them according to the new landscape requirements; (b) rational minimisation of the restriction system by territorial differentiation and specification; (c) assessment of changed recreational needs and their constructive harmonisation with environmental interests; (d) optimisation of the boundaries for the outer and inner zones of the regional park, by maximally adapting them to the natural, primarily natural complexes and territorial distribution of values, boundaries of land plots.

It is doubtful whether it is really sufficient to only indicate general principles when making decisions necessary to ensure the public interest related to essential environmental protection regimes. In such a case, the question arises whether such decisions can be considered reasonable and in accordance with the constitutional principles of the rule of law, legality, objectivity, transparency, etc. We can thus conclude that environmental issues in spatial plans are of two types: on the one hand, there are situations when environmental interest cannot be ensured, while on the other, unreasonable restrictions might be set up based on environmental issues.

3.3.7 Poland

The Polish spatial planning system is constantly criticised in the literature, due to the perceived negative role of the law in spatial planning. Pervasive legal involvement and a high degree of litigiousness have spawned various solutions. One such example is the role of administrative courts in assessing the (substantive) correctness of spatial plans. Often, however, this assessment goes beyond the legal framework.

Case Study: Court Rulings

A selection of rulings by Polish administrative courts follows in the next few paragraphs. The rulings concern complaints against spatial plans and, therefore, the assessment of plans by the courts. The rulings are made in accordance with Polish law but contain an extra-legal assessment as well.

In the first example, the court (II SA/Gd 47/20) assessed the validity of designating a service area in the local plan. The plan allowed for services in the area but imposed restrictions on the types that could be provided (limiting them to offices, catering, culture, hotels, car dealerships, health care, recreation, sports, and wellness). The court found these restrictions to be erroneous and declared the plan invalid because the present site has 'great economic value, which justifies its use for various types of services'. In addition, considering the property interests of neighbouring property owners, a broader possibility of services is justified, according to the court. Hence, the court partly based its decision (and the associated planning consequences) on an appeal to economic values of the land and the property interests of individual owners. For the court in question, these values were significantly more important than the planning restrictions imposed by the local plan.

In a second example, the court (II SA/Gd 20/20) considered a local plan that allowed for the intensification of production and service development in areas adjacent to forest areas, agricultural areas, and residential development areas. The court held that this concession significantly alters the conditions for protecting spatial order, and is therefore unacceptable, also from the perspective of neighbouring property owners. In this case, the court refers to the protection of spatial order and makes a planning assessment independently—without a robust legal basis, considering the development on a particular site contrary to the protection of spatial order.

In a third example, the court (II SA/Kr 978/20) found the provisions of the local zoning plan erroneous, which permitted the provision of telecommunication services provided by cable and underground networks. In the court's view, this was a limitation of the possibility to carry out investments in wireless telephony. Hence, the court made a subjective assessment of the type of telephone infrastructure may or may not be constructed in a particular area.

In a fourth example, the court (II SA/Kr 1301/19) declared the local plan, which provided for the productive use of land, invalid. According to the court, the critical problem is the failure to include a requirement that 'the onerous nature of the investment may not exceed the limits of the area covered by the plan'. Such conditions are contained in local plans adopted for other, neighbouring areas. Here, the court subjectively assumes that the planning restrictions for neighbouring plots should be identical.

And in a final example, the court (II SA/Sz 857/20) held that a local plan introducing a total ban on the development of agricultural land was invalid. In the court's opinion, this was not justified because, in its view, it was necessary (in the interest of the entire municipality) to build a water reservoir on the land. The court subjectively forejudges what development should be carried out on the land in question. Accordingly, the court questions the prohibition imposed on development.

In all five rulings, the courts feel competent to assess more than the compatibility of plan provisions with statutory regulations. By invalidating specific parts of plans, they comment on the very rationality of specific planning solutions. In the cases given, this even exceeds the judicially verifiable proportionality of the interference in the field of planning. In such cases, courts state that for specific reasons, the planning concept for an area should look differently, e.g., define the permissible services differently, define the location of the telecommunication network differently, or define the public interest (in the context of that development) differently.

As we have seen, the five rulings mentioned above were rather diverse. However, most of them were oriented towards expanding development possibilities in the interest of private owners. Hence, based on their subjective assessment, courts try to make a spatial adjustment. Usually, but not always, this adjustment concerns the extension of the development possibilities. In such cases, the court's assessment is detached from stronger legal foundations and is not supported by urban planning knowledge. We may surmise that it results from the specific reflections and worldview of the adjudicating judges. It boils down to an appreciation of the 'economic values' of a given space, a subjective conviction that a particular investment should be built in a given area, or a conviction that various spatial restrictions on plots located close to each other are unacceptable.

3.3.8 Romania

In Romania, the relationship between spatial planning and environmental protection is constantly shifting. There are three main reasons for it (Petrişor and Petrişor 2013): firstly, gaps exist between regulations and their enforcement. Secondly, theory and practice are usually disjointed, with theorists unable to produce planning instruments that are mature enough for use in practice, and practitioners usually averse to theory and research. And finally, current legislation is, unfortunately, scientifically uninformed, thereby running the risk of producing interventions 'with a blindfold'. Add to this that very few urban and regional plans are tendered at a fair price, and the consequence becomes apparent: Very few plans are successful in contributing to overall sustainability, as the law requires (Nowak et al. 2022).

Case Study: Background Studies, Plans, and the Question of the Environment

Procedurally, spatial plans must rely on substantiation studies, which include environmental surveys. These surveys are drafted by accredited specialists who, in addition to their primary qualification in environmental fields (ecology, geography, and environmental engineering), must prove their ability in placing knowledge in a spatial setting. This is usually achieved by attending a master's degree in planning or by building a relevant work experience (Petrişor 2010).

Note, however, that the background (or substantiation) study itself has no legally binding rules in terms of its content. Specific guidelines were drafted and approved in 1999 and 2000, for different territorial scales. Although outdated, they were nevertheless approved by the line Ministry. There are also some newer guidelines, dating back to 2006 and 2008, that were proposed, but never approved. They are also outdated. Against this background, the oldest approved guidelines are still widely used by those who need some orientation (Petrişor and Petrişor 2013). To make matters even worse, some local authorities require their usage even from specialists who can produce and use their own up-to-date methodology.

The main problem here is the huge cleavage between the outdated understanding of environmental issues, which includes the associated jargon, and the swift progress in urban ecology. Key concepts like green infrastructure, ecological corridors, or nature-based solutions are missing, with the environment still seen as a set of compartments (air, soil, and water, perhaps also fauna and flora), and not as a socio-ecological system. Moreover, although spatial planning requires interdisciplinary approaches, specialists in environmental issues are responsible only for their own background study, without usually having the chance to review other sections, even if they may ultimately generate negative environmental impacts. An example is the spatial plan for the Danube Delta, where the author of the energy chapter proposed placing wind generators within a protected area, even though their negative effects were known by the author of the environmental chapter, who was not consulted.

Moving on to the urban and regional plans themselves, their contents, as outlined by these guidelines, are comprised of two parts, i.e., the existing and proposed situation. Simple at a first glance, this requirement somewhat contradicts provisions within the environmental regulation, particularly those asking for important plans to undergo an environmental assessment. The problem is that such an assessment works with multiple scenarios, while plans work with only two scenarios: the 'do nothing' scenario, and the 'do something' scenario, i.e., the interventions proposed by the plan. The description of the current situation ('as is') includes a presentation of natural conditions (landforms, climate and climate trends, fauna, flora, and natural resources), that calls attention to natural and technological risks, environmental issues (such as sources of pollution and waste management issues), the overall quality of environment, as well as a presentation of the area's natural heritage, focused on natural protected areas, highlighting their components, and the human impacts that need to be mitigated or eliminated. However, the situation appears to be slowly changing, with recent efforts aimed at introducing ecological corridors in planning and environmental legislation, to restore the connectivity of fragmented habitats (Popescu et al. 2022).

To sum up, perhaps the main Romanian problem is that development often occurs not by rules established by General Urban Plans, but by exemptions arising from the approval of Zonal Urban Plans that contradict the provisions of the General Urban Plan. This practice is known as 'derogatory urban planning' (Hamma and Petrişor 2018), and it accounts for the massive loss and fragmentation of urban green infrastructure in all Romanian cities (Petrişor et al. 2021, 2022), as well as for the failure of planning to make a decisive contribution towards more sustainable settlements, at least from an environmental perspective.

3.4 Conclusions

Numerous similarities can be seen across CEE countries. They relate both to the characteristics of the local spatial plans and to the key problems encountered. These specific problems illustrate very clearly the challenges and dilemmas facing each national planning system.

Hence, from the weaknesses of the legal framework arise problems in Bulgaria and Poland. In Bulgaria, the legal framework was not adapted to the system's needs, thereby giving the possibility of bypassing spatial planning in many municipalities. Consequently, legal loopholes have led to numerous abuses. A similar situation exists in Poland. In this case, it is worth noting another effect: Spatial conflicts very often end up in courts, and the courts settle these conflicts based on legal criteria. Against this background, courts are forced to assess spatial conflicts, which from an urban planning perspective remain non-material. An additional problem is very well described in the Czech case study. The legislator wanted to strengthen spatial planning through a specific instrument, i.e., the regulatory plan. Unfortunately, this instrument turned out to be, thus invalidating the legislator's assumptions completely.

A separate interesting thread is the use of original national solutions. They include, for example, the Hungarian acts accelerating and facilitating the implementation of selected investments. If they bypass the basic spatial order (a situation similar to Poland), they may also exacerbate problems within spatial planning systems. The possibility of delegating the development of spatial entities, such as consultants or architectural offices, should also be carefully assessed. This is acceptable, but, as in Estonia, it must be linked to a high level of planning culture. Undoubtedly, a planning culture is enhanced by integrated development planning. The example of Latvia confirms that for this to work effectively, legal regulations alone are not enough.

Case studies on the relationship between planning, plans, and environmental and nature conservation are also interesting. Environmental and nature protection is strongly linked to the objectives of spatial planning in numerous publications. The problem, however, is that guidelines related to environmental and nature conservation are difficult to translate into the spatial planning sphere. One possible solution is to designate environmentally valuable or important zones in spatial plans, such as in Hungary. Another is setting more detailed land use parameters. In both cases, it is crucial to substantiate planning decisions with specialised studies. Note, however, that provisions of higher-level plans concerning the environment are very general and therefore difficult to implement at a lower level.

References

Act of 27 April (1976) Act on land-use planning and building order (the Building Act), 50/1976 Coll. Available online: https://www.fao.org/faolex/results/details/en/c/LEX-FAOC183199/. Accessed 31 Jan 2023

Act of 14 March (2006) Act No. 183/2006 Coll. on town and country planning and building code (Building Act). Parliament Czech Republic. https://www.mmr.cz/getmedia/9a941cf5-268b-4243-9880-d1b169fb33d6/SZ_angl.pdf. Accessed 25 Jan 2023

Action Plan for the Development of the Riga Metropolitan Area, Riga Planning Region Development Council 2020. Available online: https://rpr.gov.lv/wp-content/uploads/2021/01/Action-Plan-for-the-Development-of-the-RMA_Web-1.pdf. Accessed on 29 Dec 2022

Auziņš A, Jürgenson E, Burinskienė M (2020) Comparative analysis of spatial planning systems and practices: changes and continuity in Baltic countries. In: Methods and concepts of land management. vdf Hochschulverlag AG an der ETH Zürich, Zürich, pp 13–22. https://vdf.ch/methods-and-concepts-of-land-management-e-book.html

Auziņš A, Brokking P, Jürgenson E, Lakovskis P, Paulsson J, Romanovs A, Valčiukienė J, Viesturs J, Weninger K (2022) Land resource management policy in selected European countries. Land 11(12):2280. https://doi.org/10.3390/land11122280

Bodnár Zs (2022) Térképre tettük a Fidesz által kiemeltté nyilvánított beruházásokat az elmúlt 4 évből. https://atlatszo.hu/kozugy/2022/04/13/terkepre-tettuk-a-fidesz-altal-kiemeltte-nyilvanitott-beruhazasokat-az-elmult-4-evbol/. Accessed 04 Jan 2023

Boromisza Zs (2011) A környezetvédelem alapjai/Basics of environmental protection. Presentation. https://docplayer.hu/46102516-A-kornyezetvedelem-alapjai.html. Accessed 04 Jan 2023

Curia of Hungary (2021) "Az építésügyi hatósági eljárások gyakorlata" tárgykörben felállított joggyakorlat-elemző csoport összefoglaló véleménye/"Practice of the building authority procedures" on the practice analysis group summary opinion, 2021.El.II.JGY.K.1. https://kuria-birosag.hu/sites/default/files/joggyak/az_epitesugyi_hatosagi_eljarasok_gyakorlata_osszefoglalo_velemeny.pdf

Dimitrov S (2022) Over 50 municipalities still do not have General Development Plans, Minister Shishkov said in Burgas. Bulgarian News Agency. Available on: https://www.bta.bg/bg/news/economy/bg/367622-nad-50-obshtini-vse-oshte-nyamat-obshti-ustroystveni-planove-kaza-ministarshis?fbclid=IwAR3vQRWDFhhDSI1JE1Ks1UM3xiZ88GaGyvV2P0W9W7w9rUdsKq1yUAUDiHY. Last access on 31 Dec 2022

Dogramadjieva E (2018) The extension of Bansko ski zone: a problem or a solution. Probl Geogr J 165–181 (Bulgarian Academy of Science, Sofia) (in Bulgarian)

Dühr S, Colomb C, Nadin V (2010) European spatial planning and territorial cooperation. Routledge, London. https://doi.org/10.4324/9780203895290

Estonian Qualification Authority (2022) Occupational Qualification Certificates. Estonian Qualification Authority. https://www.kutseregister.ee/ctrl/en/Tunnistused/showKutsetunnistused/

Estonian Ministry of Justice (2013) The explanatory memorandum of the Planning Law. https://www.just.ee/sites/www.just.ee/files/elfinder/article_files/planeerimisseaduse_seletuskiri.pdf

Government Decree on the declaration of administrative authority cases related to the investment for the development of complex real estate in the XIV. District of Budapest and the realisation of related buildings and infrastructure improvements as a case of major importance from the national economic point of view and on the declaration of the investment as a case of major public interest 314/2022. (VIII. 11.). https://net.jogtar.hu/jogszabaly?docid=a2200314.kor. Accessed 04 Jan 2023

Hamma W, Petrişor A-I (2018) Urbanization and risks: case of Bejaia city in Algeria. Human Geogr 12(1):97–114

Harvey D (2005) A brief history of neoliberalism. Oxford University Press. https://doi.org/10.1093/oso/9780199283262.001.0001

Konsap A (2022) Planeerimisseaduse revisjoni tutvustus. Eesti Planeerijate Ühing

Naydenov N (2022) The expansion of urbanized areas—development or cline. Annual of the University of Architecture, Civil Engineering and Geodesy, Sofia (in Press) (in Bulgarian)
Nowak MJ, Petrişor A-I, Petrişor A-I, Filepné KK, Lukstina G, Jürgenson E, Ladzianska Z, Simeonova V, Lozyskyy R, Rezac V, Pantyley V, Praneviciene B, Fakeyeva L, Mickiewicz B, Blaszke M (2022) The role of spatial plans adopted at the local level in the spatial planning systems of Central and Eastern European countries. Land 11(9):1599
Official Site of the National Register of Buildings. Provisions of building services. https://www.e-epites.hu/hatosag/epitesugyi-feladatellatas. Accessed 04 Jan 2023
Official Website of District Újbuda, Budapest (2022) A varázsige: kiemelt beruházás. https://www.ujbuda.hu/ujbuda/a-varazsige-kiemelt-beruhazas. Accessed 04 Jan 2023
Parliament of Estonia (Riigikogu). General part of the Environmental Code Act. Passed 16.02.2011. https://www.riigiteataja.ee/en/eli/517062022003/consolide
Parliament of Estonia (Riigikogu). Planning Act. Passed 28.01.2015. https://www.riigiteataja.ee/en/eli/ee/Riigikogu/act/525102022002/consolide
Passed by referendum. The Constitution of the Republic of Estonia. Passed 28.06.1992. https://www.riigiteataja.ee/en/eli/530122020003/consolide
Petrişor A-I (2010) The theory and practice of urban and spatial planning in Romania: education, laws, actors, procedures, documents, plans, and spatial organization. A multiscale analysis. Serb Architectural J 2(2):139–154
Petrişor A-I, Petrişor LE (2013) The shifting relationship between urban and spatial planning and the protection of the environment: Romania as a case study. Present Environ Sustain Dev 7(1):268–276
Petrişor A-I, Mierzejewska L, Mitrea A, Drachal K, Tache AV (2021) Dynamics of open green areas in Polish and Romanian cities during 2006–2018: insights for spatial planners. Remote Sens 13(20):4041
Petrişor A-I, Mierzejewska L, Mitrea A (2022) Mechanisms of change in urban green infrastructure—evidence from Romania and Poland. Land 11(5):592
Popescu OC, Tache AV, Petrişor A-I (2022) Methodology for identifying ecological corridors: a spatial planning perspective. Land 11(7):1013
Riga Planning Region Sustainable Development Strategy 2014–2030 (update) (1.0 version). Riga Planning Region Development Council 2022. Available online: https://geolatvija.lv/geo/tapis#document_20758. Accessed on 29 Dec 2022
Szabó Z, Balas G, Borbas G, Kiss G, Piross A, Potvorszki G, Szerletics Á (2016) Európai Uniós források rendszere 2014–2020 Gyakorlati útmutató pályázóknak és pályáztatóknak
Tájoló-terv Kft. (2017) Környezeti értékelés, Szentendre Város, településrendezési eszközök. Environmental assessment, spatial planning tools, Szentendre. Available online: https://szentendre.hu/wp-content/uploads/2017/02/DOKU_SZE-Kornyezeti-ertekeles.pdf. Accessed 04 Jan 2023
Tartu City Government (Tartu Linnavalitsus) (2022) Comprehensive plan Tartu 2040+. https://gis.tartulv.ee/yldplaneering2040/
Vodenska M (1992) International tourism in Bulgaria: problems and perspectives. In: Tijdschrift voor Economische en Sociale Geografie, vol 83, no 5, pp 409–417, ref 7

The manufacturer's authorised representative in the EU is Springer Nature Customer Service Centre GmbH, Europaplatz 3, 69115 Heidelberg, Germany. If you have any concerns regarding our products, please contact ProductSafety@springernature.com

Printed and bound by CPI Group (UK) Ltd, Croydon, CR0 4YY
01/10/2026
02225918-0001